GENERAL HOMOGENEOUS COORDINATES IN SPACE OF THREE DIMENSIONS

GENERAL HOMOGENEOUS COORDINATES IN SPACE OF THREE DIMENSIONS

BY

E. A. MAXWELL

Fellow of Queens' College, Cambridge

CAMBRIDGE

AT THE UNIVERSITY PRESS

1959

CAMBRIDGE UNIVERSITY PRESS
Cambridge, New York, Melbourne, Madrid, Cape Town, Singapore, São Paulo, Delhi

Cambridge University Press
The Edinburgh Building, Cambridge CB2 8RU, UK

Published in the United States of America by Cambridge University Press, New York

www.cambridge.org
Information on this title: www.cambridge.org/9780521092289

First published 1951
Reprinted 1959
Re-issued in this digitally printed version 2008

A catalogue record for this publication is available from the British Library

ISBN 978-0-521-09228-9 paperback

DEDICATED

TO

MY WIFE

GRETA LOUISE MAXWELL

CONTENTS

PREFACE

I AM deeply indebted to two lecturers in the University of Cambridge for their help and advice in the preparation of this book. The manuscript was read by Dr S. Wylie and the proofs by Dr J. A. Todd, F.R.S., and the adoption of their suggestions has added considerably to the clarity and accuracy of the text.

I should also like to record my thanks to a number of pupils, whose corrections I was happy to receive.

To the staff of the Cambridge University Press I again express my appreciation of their printing and of the courteous help which I have always received.

E.A.M.

QUEENS' COLLEGE
CAMBRIDGE

December 1950

INTRODUCTION

THE purpose of this book is at once modest and ambitious, namely, to provide a *short* introduction to algebraic geometry in space of three dimensions, to make clear its spirit, and to prepare the way for deeper study. I have in mind a reader who has just read my book on homogeneous coordinates in a plane (to which this stands as a second volume) and is in the early stages of his second-year work at the University. I have also in mind a class of reader who has read further in mathematics generally, but has found the existing detailed accounts of this work too full or too specialised for his own needs; I hope that such a reader will find here a temptation to get to grips with the subject.

In spite of the existence of a large number of text-books on the geometry of space of three dimensions, I think it is true that there are few which deal with the subject in the essential spirit of projective geometry. The two accounts which seem to me to be of greatest importance for further study are, first, a well-established authority, the *Principles of Geometry*, Vol. III, by Prof. H. F. Baker, and, secondly, an important recent work, *Projective and Analytical Geometry*, by Dr J. A. Todd which for the first time (I think) establishes in readily available form the synthesis of projective geometry with modern matrix algebra. The aim of the present book will be fulfilled if it encourages the reader to turn to these two accounts and, perhaps, helps him a little along the way.

In place of the elementary examples which usually appear throughout each chapter in a book of this kind, I have included a fairly large number of *Theorem-examples*; these are almost entirely standard results, which should be known, and which follow directly from the preceding work. *The conscientious solution of these examples is an essential part of the reading of this book*, and I have added short hints which should remove any possible difficulty.

A reader at this stage will almost certainly be under a teacher, and may well have to call upon him for help in the solution of some of the Miscellaneous Examples at the ends of the chapters. These have appeared in examination papers at what we now seem to call 'the highest level', and are not all easy. They are taken from Preliminary (P.) or Mathematical Tripos (M.T.) papers set in the University of Cambridge, or from Honours papers set in the University of London (L.), and I am grateful for permission to use them. In the earlier chapters I have preceded the Miscellaneous Examples with some numerical work to give practice in manipulation.

There are occasional references to my earlier book, which for brevity I denote by the letter M.

Finally, a word should be said about the last chapter. I have thought it right to bring the reader to the threshold of the methods now in use, but appreciate that algebraic equipment may vary considerably. I have therefore given a brief self-contained summary of the elements of matrix algebra, and then demonstrated how it can be applied. The reader will readily appreciate the mental economy which the introduction of matrices provides. As far as I am able to judge, the paragraph on line-coordinates in this chapter contains several results which simplify existing treatments.

E. A. M.

CHAPTER I

THE POINT, THE STRAIGHT LINE AND THE PLANE

1. Knowledge assumed. This book is an immediate sequel to the author's earlier volume *The Methods of Plane Projective Geometry based on the use of General Homogeneous Coordinates* (Cambridge University Press, 1946), and knowledge of the subject-matter will be assumed. We also quote certain obvious extensions to space of three dimensions without giving detailed proofs.

As in the book just named, we again confine our first attention to projective geometry, in which the ideas of length and angle are not used. There are now four homogeneous coordinates instead of three, not necessarily real. The basis on which the use of complex coordinates rests is closely analogous to the development given for plane geometry, but for a rigorous treatment the reader should consult a more advanced text-book.

2. The homogeneous coordinates. We make the following basic assumption:

The position of a point can be uniquely defined by the ratios of four coordinates x, y, z, t, and, conversely, these ratios define a point uniquely.

We assume that x, y, z, t are not all zero simultaneously. Apart from that, they may take any sets of values, real or complex. In speaking of the point with coordinates x, y, z, t we shall often call it simply 'the point (x, y, z, t)'; if we have given the point a name, say P, we shall speak of it as '$P(x, y, z, t)$'.

3. The symbol of a point. It is sometimes convenient to have a single composite sign for the coordinates (x, y, z, t) of a point P, and we use the notation **P** for this purpose.* We call **P** the *symbol* of the point $P(x, y, z, t)$, and sometimes speak of '**P**' as a point, meaning 'the point P whose symbol is **P**'.

* The reader who is familiar with matrices may read **P** as denoting the column matrix $\{x, y, z, t\}$. This idea will be developed in Chapter VIII.

If $P_1, P_2, \ldots, P_n$ are n given points, and if non-zero numbers $\lambda_1, \lambda_2, \ldots, \lambda_n$ can be found such that

$$\lambda_1 \mathbf{P}_1 + \lambda_2 \mathbf{P}_2 + \ldots + \lambda_n \mathbf{P}_n \equiv 0,$$

by which we mean that there exist relations

$$\begin{aligned} \lambda_1 x_1 + \lambda_2 x_2 + \ldots + \lambda_n x_n &= 0, \\ \lambda_1 y_1 + \lambda_2 y_2 + \ldots + \lambda_n y_n &= 0, \\ \lambda_1 z_1 + \lambda_2 z_2 + \ldots + \lambda_n z_n &= 0, \\ \lambda_1 t_1 + \lambda_2 t_2 + \ldots + \lambda_n t_n &= 0 \end{aligned}$$

connecting the coordinates, then we say that the points

$$P_1, P_2, \ldots, P_n$$

are *in syzygy*; the identity itself is called a *syzygy*.

If $n = 2$, the syzygy between P_1 and P_2 implies that they are in fact the same point. We shall always assume that no two points are in syzygy unless the contrary possibility is stated explicitly.

4. The straight line. Let $A(x_1, y_1, z_1, t_1)$, $B(x_2, y_2, z_2, t_2)$ be two given points. *We define the line AB to consist of the points*

$$P(x, y, z, t)$$

for which values of the ratio $\lambda:\mu$ *can be found such that*

$$\begin{aligned} x &= \lambda x_1 + \mu x_2, \\ y &= \lambda y_1 + \mu y_2, \\ z &= \lambda z_1 + \mu z_2, \\ t &= \lambda t_1 + \mu t_2. \end{aligned}$$

Every value of $\lambda:\mu$ determines one and only one point,* which is called *a point of the line AB*. In particular, $\mu = 0$ determines A and $\lambda = 0$ determines B. The two values λ, μ cannot vanish simultaneously.

* It is, of course, understood that the coordinates (x_1, y_1, z_1, t_1), (x_2, y_2, z_2, t_2), and not merely the ratios, are completely settled before applying the definition.

The four equations just given can be expressed compactly in terms of the symbols, in the form

$$\mathbf{P} \equiv \lambda\mathbf{A} + \mu\mathbf{B}.$$

The identity between the symbols of three points on a straight line is thus a syzygy. More symmetrically, if the symbols **A**, **B**, **C** of three points A, B, C are connected by means of the syzygy

$$\lambda\mathbf{A} + \mu\mathbf{B} + \nu\mathbf{C} \equiv 0,$$

then each point lies on the line joining the other two, and the three points are collinear.

Theorem-examples.* 1. A straight line is determined by *any* two of its points. [Compare M., p. 5.]

2. If the symbols of four points A, B, C, D are connected by means of a syzygy

$$\lambda\mathbf{A} + \mu\mathbf{B} + \nu\mathbf{C} + \rho\mathbf{D} \equiv 0,$$

then the lines AD, BC intersect (or, in a special case, coincide).

[The symbol of their common point is a multiple of either $\lambda\mathbf{A} + \rho\mathbf{D}$ or $\mu\mathbf{B} + \nu\mathbf{C}$.]

3. The two points whose symbols are $\lambda\mathbf{A} \pm \mu\mathbf{B}$ separate A, B harmonically.

[The parameter λ/μ in the symbol $\lambda\mathbf{A} + \mu\mathbf{B}$ determines the individual points of the line, and the cross-ratio of four points is the cross-ratio of the four corresponding parameters.]

4. If O, A, B, C are four given points and A', B', C' three given points on OA, OB, OC respectively, then, by adjusting multiples of **A**, **B**, **C** when necessary, the symbols of A', B', C' can be taken, without loss of generality, in the respective forms

$$\mathbf{O} + \mathbf{A}, \quad \mathbf{O} + \mathbf{B}, \quad \mathbf{O} + \mathbf{C}.$$

5. The plane. Let

$$A(x_1, y_1, z_1, t_1), \quad B(x_2, y_2, z_2, t_2), \quad C(x_3, y_3, z_3, t_3)$$

be three given NON-COLLINEAR points. *We define the plane ABC to consist of the points $P(x, y, z, t)$ for which values of the ratios $\lambda : \mu : \nu$ can be found such that*

$$x = \lambda x_1 + \mu x_2 + \nu x_3,$$
$$y = \lambda y_1 + \mu y_2 + \nu y_3,$$
$$z = \lambda z_1 + \mu z_2 + \nu z_3,$$
$$t = \lambda t_1 + \mu t_2 + \nu t_3.$$

* The Theorem-examples are a basic part of the text, and should be both solved and remembered.

Every set of values of $\lambda:\mu:\nu$ (not all zero) determines one and only one point, which is called *a point of the plane ABC*. In particular, $\mu = \nu = 0$ determines A, $\nu = \lambda = 0$ determines B and $\lambda = \mu = 0$ determines C. The three values λ, μ, ν cannot vanish simultaneously.

The four equations just given can be expressed compactly in terms of the symbols, by means of the syzygy

$$\mathbf{P} \equiv \lambda\mathbf{A} + \mu\mathbf{B} + \nu\mathbf{C}.$$

The coordinates of P satisfy the relation, found by eliminating $-1:\lambda:\mu:\nu$,

$$\begin{vmatrix} x & x_1 & x_2 & x_3 \\ y & y_1 & y_2 & y_3 \\ z & z_1 & z_2 & z_3 \\ t & t_1 & t_2 & t_3 \end{vmatrix} = 0.$$

This relation, on expansion, assumes the form

$$lx + my + nz + pt = 0,$$

where l, m, n, p are the cofactors of x, y, z, t in the determinant. This is a homogeneous linear equation in the four variables x, y, z, t, and is called the *equation of the plane.*

Conversely, every point $P(x, y, z, t)$ whose coordinates satisfy the relation

$$\begin{vmatrix} x & x_1 & x_2 & x_3 \\ y & y_1 & y_2 & y_3 \\ z & z_1 & z_2 & z_3 \\ t & t_1 & t_2 & t_3 \end{vmatrix} = 0$$

does lie in the plane ABC. For there then exist numbers p, q, r, s such that

$$px + qx_1 + rx_2 + sx_3 = 0, \text{ etc.};$$

moreover, p cannot be zero, otherwise the points **A**, **B**, **C** would be connected by a syzygy $q\mathbf{A} + r\mathbf{B} + s\mathbf{C} \equiv \mathbf{0}$ and so be collinear, contrary to the hypothesis. Dividing by p and writing $q/p = -\lambda$, $r/p = -\mu$, $s/p = -\nu$, we find the relation expressed concisely in the form

$$\mathbf{P} \equiv \lambda\mathbf{A} + \mu\mathbf{B} + \nu\mathbf{C},$$

which shows, by definition, that P lies in the plane ABC.

It is customary to interchange rows and columns and to write the *equation of the plane ABC* in the form

$$\begin{vmatrix} x & y & z & t \\ x_1 & y_1 & z_1 & t_1 \\ x_2 & y_2 & z_2 & t_2 \\ x_3 & y_3 & z_3 & t_3 \end{vmatrix} = 0.$$

An immediate corollary of the preceding work is that, *if the points (x_i, y_i, z_i, t_i), with $i = 1, 2, 3, 4$, are coplanar, then*

$$\begin{vmatrix} x_1 & y_1 & z_1 & t_1 \\ x_2 & y_2 & z_2 & t_2 \\ x_3 & y_3 & z_3 & t_3 \\ x_4 & y_4 & z_4 & t_4 \end{vmatrix} = 0.$$

Conversely, *if this determinant vanishes, then the points are coplanar.*

Theorem-examples. 1. If the symbols of four distinct points A, B, C, D are connected by the syzygy

$$\lambda \mathbf{A} + \mu \mathbf{B} + \nu \mathbf{C} + \rho \mathbf{D} \equiv 0,$$

then each point lies in the plane determined by the other three, and the four points are coplanar.

How is this result modified if **A**, **B**, **C** are themselves connected by a syzygy?

2. A plane is determined by *any* three of its points which are not collinear. [Use an argument similar to M., p. 5.]

3. Every linear equation

$$lx + my + nz + pt = 0,$$

where l, m, n, p are not all zero, does determine a plane, and determines it uniquely.

[Three points of the plane are $(-p, 0, 0, l)$, $(0, -p, 0, m)$, $(0, 0, -p, n)$. Compare M., p. 6.]

4. The equations

$$l_1 x + m_1 y + n_1 z + p_1 t = 0, \quad l_2 x + m_2 y + n_2 z + p_2 t = 0$$

determine the same plane if $l_1 : l_2 = m_1 : m_2 = n_1 : n_2 = p_1 : p_2$.

5. A *necessary and sufficient* condition that the four planes

$$l_i x + m_i y + n_i z + p_i t = 0 \quad (i = 1, 2, 3, 4)$$

have a common point is

$$\begin{vmatrix} l_1 & m_1 & n_1 & p_1 \\ l_2 & m_2 & n_2 & p_2 \\ l_3 & m_3 & n_3 & p_3 \\ l_4 & m_4 & n_4 & p_4 \end{vmatrix} = 0.$$

[Compare M., p. 7.]

6. The intersection of two planes. Let the equations of two given distinct planes be

$$l_1 x + m_1 y + n_1 z + p_1 t = 0,$$
$$l_2 x + m_2 y + n_2 z + p_2 t = 0.$$

Since the planes are distinct, the coefficients $l_1, m_1, n_1, p_1; l_2, m_2, n_2, p_2$ are not proportional, and so we may assume, without loss of generality, that $l_1 p_2 - l_2 p_1$ does not vanish. By putting $y = 0$, and then solving the two equations, we see that the point

$$(n_1 p_2 - n_2 p_1, \quad 0, \quad p_1 l_2 - p_2 l_1, \quad l_1 n_2 - l_2 n_1)$$

lies in each plane; and, by putting $z = 0$, we see similarly that the point

$$(m_1 p_2 - m_2 p_1, \quad p_1 l_2 - p_2 l_1, \quad 0, \quad l_1 m_2 - l_2 m_1)$$

also lies in each plane. [The hypothesis $l_1 p_2 - l_2 p_1 \neq 0$ ensures that these coordinates do not all vanish.] Hence there are certainly two distinct points, which it will be convenient to call $P_1(x_1, y_1, z_1, t_1)$, $P_2(x_2, y_2, z_2, t_2)$, common to both planes. But then the coordinates of the point

$$P(\lambda x_1 + \mu x_2, \lambda y_1 + \mu y_2, \lambda z_1 + \mu z_2, \lambda t_1 + \mu t_2)$$

satisfy each of the two given equations, so that P lies in each of the given planes. Further, no point outside the line $P_1 P_2$ can be common to both planes; for if there were such a point, say P_3, then each of the given planes would coincide with the plane $P_1 P_2 P_3$, contrary to the hypothesis that they are distinct. Hence *the points common to two given planes lie in a straight line.* This line is called the *line of intersection* of the two planes.

If the two given planes are called π_1, π_2, then we denote their line of intersection by the symbol $(\pi_1 \pi_2)$.

A straight line is determined when two distinct planes through it are given, and so the coordinates of the points of a line satisfy each of *two* linear equations, say

$$l_1x+m_1y+n_1z+p_1t = 0,$$
$$l_2x+m_2y+n_2z+p_2t = 0.$$

These equations are called the *equations of the line*. The two equations are not themselves uniquely determined; either may be replaced by an equation formed by adding any multiple of the first equation to any multiple of the second. For example, the line whose equations are $x = t = 0$ is equally well defined by the equations

$$ax+bt = 0, \quad a'x+b't = 0,$$

where $a/a' \neq b/b'$.

Theorem-examples. 1. The plane

$$\lambda(l_1x+m_1y+n_1z+p_1t)+\mu(l_2x+m_2y+n_2z+p_2t) = 0$$

passes through the line of intersection of the two planes

$$l_ix+m_iy+n_iz+p_it = 0 \quad (i = 1, 2).$$

Conversely, the equation of *any* plane through the line can be expressed in that form.

[Compare M., pp. 13–14.]

2. Two straight lines which lie in a plane have one point in common. If l_1, l_2 are the lines, we call the point $(l_1 l_2)$.

3. The line joining two points in a plane lies entirely in that plane. If A_1, A_2 are the points, we call the line A_1A_2.

4. A line not lying in a plane meets it in one, and only one, point. If l is the line and π the plane, we call the point $(l\pi)$, or (πl).

5. A unique plane can be drawn through a given point to contain a given line not passing through that point. If P is the point and l the line, we call the plane $[Pl]$, or $[lP]$.

6. A unique plane can be drawn to contain two given lines with one point in common. If l_1, l_2 are the lines, we call the plane $[l_1 l_2]$.

7. Two lines which do not lie in a plane have no common point, and two lines which have no common point do not lie in a plane. Two such lines are said to be *skew*.

8. Three distinct planes either have a line in common or meet in a unique point. In the latter case, if π_1, π_2, π_3 are the planes, we call the point $(\pi_1\pi_2\pi_3)$.

9. Three lines, each of which meets the other two, either lie in a plane or meet in a point.

10. The four planes whose equations are

$$\begin{aligned} -ny + mz - l't &= 0,\\ nx - lz - m't &= 0,\\ -mx + ly - n't &= 0,\\ l'x + m'y + n'z &= 0, \end{aligned}$$

where the numbers l, m, n, l', m', n' are connected by the relation

$$ll' + mm' + nn' = 0,$$

have a line in common.

[Consider $n'(2) - m'(3) + l(4)$, etc.]

DEFINITION. The system of planes passing through a given line l is called a *pencil*. Thus if l is defined by the planes π_1, π_2 whose equations are

$$\pi_1 \equiv l_1x + m_1y + n_1z + p_1t = 0,$$
$$\pi_2 \equiv l_2x + m_2y + n_2z + p_2t = 0,$$

then the equation of any plane of the pencil is

$$\pi_1 + \lambda\pi_2 = 0.$$

The line l is called the *axis* of the pencil.

In the same way, the system of points lying on a given line l is called a *range*. Thus if l is defined by the points $P_1(x_1, y_1, z_1, t_1)$, $P_2(x_2, y_2, z_2, t_2)$, then the coordinates of any point of the range are

$$(x_1 + \lambda x_2, y_1 + \lambda y_2, z_1 + \lambda z_2, t_1 + \lambda t_2).$$

The line l is called the *base* of the range.

7. The syzygy connecting the symbols of five given points. Let P_1, P_2, P_3, P_4, P_5 be five given points, of which no four are coplanar. We prove that *their symbols are necessarily connected by a syzygy, uniquely defined by* $\mathbf{P}_1, \mathbf{P}_2, \mathbf{P}_3, \mathbf{P}_4, \mathbf{P}_5$, *namely*

$$\lambda_1\mathbf{P}_1 + \lambda_2\mathbf{P}_2 + \lambda_3\mathbf{P}_3 + \lambda_4\mathbf{P}_4 + \lambda_5\mathbf{P}_5 \equiv \mathbf{0}.$$

We have to prove that the four equations

$$\begin{aligned} \lambda_1x_1 + \lambda_2x_2 + \lambda_3x_3 + \lambda_4x_4 + \lambda_5x_5 &= 0,\\ \lambda_1y_1 + \lambda_2y_2 + \lambda_3y_3 + \lambda_4y_4 + \lambda_5y_5 &= 0,\\ \lambda_1z_1 + \lambda_2z_2 + \lambda_3z_3 + \lambda_4z_4 + \lambda_5z_5 &= 0,\\ \lambda_1t_1 + \lambda_2t_2 + \lambda_3t_3 + \lambda_4t_4 + \lambda_5t_5 &= 0 \end{aligned}$$

can be solved uniquely for the ratios $\lambda_1:\lambda_2:\lambda_3:\lambda_4:\lambda_5$. It is well known that, if D_i denotes the determinant

$$D_i \equiv \begin{vmatrix} x_j & x_k & x_l & x_m \\ y_j & y_k & y_l & y_m \\ z_j & z_k & z_l & z_m \\ t_j & t_k & t_l & t_m \end{vmatrix},$$

where i, j, k, l, m are the numbers 1, 2, 3, 4, 5 in cyclic order, then

$$\frac{\lambda_1}{D_1} = \frac{\lambda_2}{D_2} = \frac{\lambda_3}{D_3} = \frac{\lambda_4}{D_4} = \frac{\lambda_5}{D_5},$$

where the determinants D_i are not zero, since no three points are collinear or four points coplanar.* The syzygy is therefore uniquely determined.

Theorem-example. The symbol of a point P in general position in space can be expressed in terms of those of four given points X, Y, Z, T (not coplanar) by means of a syzygy

$$\mathbf{P} \equiv \lambda\mathbf{X} + \mu\mathbf{Y} + \nu\mathbf{Z} + \rho\mathbf{T}.$$

8. The transversal from a given point to two given skew lines. Let P be a given point and l_1, l_2 two given skew lines not through P. The planes $[Pl_1]$, $[Pl_2]$ are uniquely determined (§ 6, Theorem-example 5), and they meet in a line, necessarily through P. This line meets l_1, l_2, since it lies in the planes $[Pl_1]$, $[Pl_2]$ respectively. It is called the *transversal* from P to the given lines.

Alternatively, let P be the given point and A_1B_1, A_2B_2 the two given skew lines. The five symbols $\mathbf{P}$, $\mathbf{A}_1$, $\mathbf{B}_1$, $\mathbf{A}_2$, $\mathbf{B}_2$ are connected by a syzygy, say

$$\lambda\mathbf{P} + \mu_1\mathbf{A}_1 + \nu_1\mathbf{B}_1 + \mu_2\mathbf{A}_2 + \nu_2\mathbf{B}_2 \equiv \mathbf{0},$$

which we can arrange in the form

$$\lambda\mathbf{P} + (\mu_1\mathbf{A}_1 + \nu_1\mathbf{B}_1) + (\mu_2\mathbf{A}_2 + \nu_2\mathbf{B}_2) \equiv \mathbf{0}.$$

This relation shows that the points with symbols $\mathbf{P}$, $\mu_1\mathbf{A}_1 + \nu_1\mathbf{B}_1$, $\mu_2\mathbf{A}_2 + \nu_2\mathbf{B}_2$ are collinear, as required.

* If three of the points were collinear, then one of the columns would be the sum of appropriate multiples of two others.

Theorem-example. If P lies on l_1, there is an infinity of lines meeting l_1 and l_2.

[This is, of course, obvious; but the reader should consider what happens to the syzygy in the alternative proof.]

9. The tetrahedron of reference.

DEFINITION. The figure formed by four non-concurrent planes is called a *tetrahedron*. The planes are called the *faces* of the tetrahedron, the four points in which three of the planes meet are called the *vertices*, and the six lines in which the planes meet in pairs are called the *edges*. Two vertices lie on each edge.

In particular, the four points $X(1, 0, 0, 0)$, $Y(0, 1, 0, 0)$, $Z(0, 0, 1, 0)$, $T(0, 0, 0, 1)$ are the vertices of a tetrahedron called the *tetrahedron of reference*. The faces YZT, ZXT, XYT, XYZ are given by the equations

$$x = 0, \quad y = 0, \quad z = 0, \quad t = 0$$

respectively, and the edges YZ, ZX, XY, XT, YT, ZT are given by the pairs of equations

$$x = t = 0; \quad y = t = 0; \quad z = t = 0;$$

$$y = z = 0; \quad z = x = 0; \quad x = y = 0$$

respectively.

10. The unit point and the unit plane. If the tetrahedron of reference is given, we can refer to it a system of coordinates in which any assigned point U, not on a face of the tetrahedron, has coordinates $(1, 1, 1, 1)$, as follows:

Suppose that, in *any* coordinate system with that tetrahedron of reference, the coordinates of U are $(\alpha, \beta, \gamma, \delta)$. We can, as it were, 're-name' the points of space, by means of a *transformation* from the system x, y, z, t to a system x', y', z', t' given by the relations

$$x' = x/\alpha, \quad y' = y/\beta, \quad z' = z/\gamma, \quad t' = t/\delta.$$

Then (i) the coordinates of every point of space are determined in terms of x', y', z', t'; (ii) the point $(\alpha, \beta, \gamma, \delta)$ becomes the point $(1, 1, 1, 1)$; (iii) the tetrahedron of reference is unchanged. We have therefore found a system of coordinates in which U is the point $(1, 1, 1, 1)$, called the *unit point* for that system of coordinates.

We may similarly simplify the equation of the plane

$$lx+my+nz+pt = 0$$

to the form

$$x+y+z+t = 0$$

(the *unit plane*) by means of the transformation

$$x' = lx, \quad y' = my, \quad z' = nz, \quad t' = pt.$$

Note that *the two simplifications, unit point and unit plane, cannot be effected simultaneously for an arbitrary point as well as for an arbitrary plane*. (See Illustration 2, p. 17.)

Theorem-examples. 1. The syzygy connecting the symbols of five points in general position in space can be expressed in the form

$$\mathbf{A}+\mathbf{B}+\mathbf{C}+\mathbf{D}+\mathbf{E}\equiv 0.$$

2. The syzygies connecting four general points in a plane or three distinct points in a line can be expressed respectively in the forms

$$\mathbf{A}+\mathbf{B}+\mathbf{C}+\mathbf{D}\equiv 0,$$
$$\mathbf{A}+\mathbf{B}+\mathbf{C}\equiv 0.$$

11. Duality. The concept of duality in space is naturally a little more elaborate than in a plane, but the reader who has grasped the latter should have little further difficulty. We find, in fact, that *point* and *plane* are dual conceptions, but that, in addition, the line is a self-dual construct in the sense that the system of points on a line is dual to the system of planes through a line. The reader should compare the following statements, which we (with his help, we hope) have proved earlier:

{Three points determine a plane;
Three planes determine a point.

{Two points determine a line;
Two planes determine a line.

{Two lines in a plane meet in a point;
Two lines through a point lie in a plane.

{A line meets a plane in a point;
A line and a point determine a plane.

If we have a figure consisting of a number of points, lines and planes, we can form a *dual* figure as follows:

Replace every point A, B, C, ... by a plane α', β', γ', ...;

Replace every plane ξ, η, ζ, ... by a point X', Y', Z', ...;

Replace every line l, m, n, ... by a line l', m', n', ...;

Replace every line AB by a line $(\alpha'\beta')$ and every line $(\xi\eta)$ by a line $X'Y'$;

Replace every point $(l\xi)$ by a plane $[l'X']$ and every plane $[lA]$ by a point $(l'\alpha')$;

Replace every point $(\xi\eta\zeta)$ by a plane $[X'Y'Z']$ and every plane $[ABC]$ by a point $(\alpha'\beta'\gamma')$.

Theorem-examples. 1. The (space) dual of a triangle ABC is the figure of three planes meeting in a point.

2. The tetrahedron is (in a sense to be stated explicitly) a self-dual construct.

12. Plane coordinates. The idea of duality finds its exact interpretation in terms of the coordinates. We have seen that the equation of any plane can be expressed in the form

$$\pi \equiv lx + my + nz + pt = 0.$$

In other words, the plane π is determined when the ratios $l:m:n:p$ are given; conversely, the ratios can be determined when the plane is given. We may therefore use l, m, n, p as a system of *coordinates* to determine the plane. We call them *plane coordinates*, or *tangential coordinates*, and we refer to the plane π whose coordinates are l, m, n, p as 'the plane (l, m, n, p)', or 'the plane π (l, m, n, p)'. When the distinction is necessary, we shall refer to x, y, z, t as *point-coordinates*.

13. The equation of a point. The equation

$$ax + by + cz + dt = 0$$

asserts that every point (x, y, z, t) subject to it lies in the plane whose (plane) coordinates are (a, b, c, d).

Dually, the equation

$$al + bm + cn + dp = 0$$

in plane coordinates asserts that every plane (l, m, n, p) subject to it contains the point whose point-coordinates are (a, b, c, d).

Now put the two results together:

(i) The equation

$$ax + by + cz + dt = 0$$

in point-coordinates (x, y, z, t) represents the points of a plane, and the plane coordinates of that plane are (a, b, c, d). We speak of the equation as *the* (*point*) *equation of the plane.*

(ii) The equation

$$al + bm + cn + dp = 0$$

in plane coordinates (l, m, n, p) represents the planes through a point, and the point-coordinates of that point are (a, b, c, d). We speak of the equation as *the equation of the point.*

14. Transformation of coordinates. We have already seen how to make a change of coordinates in which the tetrahedron of reference is unaltered. Let us now consider the effect of the transformation

$$\left.\begin{aligned} \xi &= a_1x + b_1y + c_1z + d_1t, \\ \eta &= a_2x + b_2y + c_2z + d_2t, \\ \zeta &= a_3x + b_3y + c_3z + d_3t, \\ \tau &= a_4x + b_4y + c_4z + d_4t. \end{aligned}\right\} \qquad (1)$$

If, as we assume, the determinant

$$\Delta \equiv \begin{vmatrix} a_1 & b_1 & c_1 & d_1 \\ a_2 & b_2 & c_2 & d_2 \\ a_3 & b_3 & c_3 & d_3 \\ a_4 & b_4 & c_4 & d_4 \end{vmatrix}$$

is not zero, then we can solve these equations for x, y, z, t (by multiplying by appropriate cofactors and adding) to obtain relations which we write in the form

$$\left.\begin{aligned} \Delta x &= A_1\xi + A_2\eta + A_3\zeta + A_4\tau, \\ \Delta y &= B_1\xi + B_2\eta + B_3\zeta + B_4\tau, \\ \Delta z &= C_1\xi + C_2\eta + C_3\zeta + C_4\tau, \\ \Delta t &= D_1\xi + D_2\eta + D_3\zeta + D_4\tau, \end{aligned}\right\} \qquad (2)$$

where $A_1, A_2, \ldots$ are the cofactors of $a_1, a_2, \ldots$ in the determinant Δ.

With the assumption $\Delta \neq 0$, the four planes

$$\xi = 0, \quad \eta = 0, \quad \zeta = 0, \quad \tau = 0$$

form a tetrahedron, since they are not concurrent. We denote the point $\eta = \zeta = \tau = 0$ by U, $\zeta = \xi = \tau = 0$ by V, $\xi = \eta = \tau = 0$ by W, $\xi = \eta = \zeta = 0$ by P. When we are given any configuration specified by coordinates x, y, z, t, the equations (2) enable us to specify the configuration alternatively in terms of ξ, η, ζ, τ, and the new system of coordinates then has the tetrahedron of reference $UVWP$.

It follows that *we can select any tetrahedron to be the tetrahedron of reference* for the system of coordinates ξ, η, ζ, τ. Having done so, we may also choose an assigned point as the new 'unit' point, just as in § 10.

The fundamental property of this transformation is that *a homogeneous equation of degree n in x, y, z, t is transformed into a homogeneous equation of degree n in ξ, η, ζ, τ.* In particular, a plane is transformed into a plane; in fact, the plane

$$lx + my + nz + pt = 0$$

is transformed into the plane

$$l(A_1\xi + A_2\eta + A_3\zeta + A_4\tau) + m(B_1\xi + B_2\eta + B_3\zeta + B_4\tau) + n(C_1\xi + C_2\eta + C_3\zeta + C_4\tau) + p(D_1\xi + D_2\eta + D_3\zeta + D_4\tau) = 0.$$

Hence the corresponding law of transformation for the plane coordinates is

$$\begin{aligned} l' &= A_1 l + B_1 m + C_1 n + D_1 p, \\ m' &= A_2 l + B_2 m + C_2 n + D_2 p, \\ n' &= A_3 l + B_3 m + C_3 n + D_3 p, \\ p' &= A_4 l + B_4 m + C_4 n + D_4 p. \end{aligned}$$

15. Cross-ratio properties. Just as in plane geometry, we can also consider here the cross-ratio of four elements enumerated by means of an algebraic parameter. For example, we have the cross-ratio of four points on a line, of four coplanar lines through a point, and of four planes through a line. The properties are very similar to those in a plane, and we do not propose to go into much detail. The reader may refresh his memory by consulting M., Chapters II and III.

Theorem-examples. 1. Four given planes of a pencil are met by an arbitrary line in four points whose cross-ratio is constant. This cross-ratio may be called *the cross-ratio of the four planes of the pencil.*

2. The four lines which join an arbitrary point of space to four given collinear points are cut by an arbitrary plane in four collinear points whose cross-ratio is constant.

16. A transversal theorem for a tetrahedron (Von Staudt's theorem). *A line l cuts the faces YZT, ZXT, XYT, XYZ in points A, B, C, D. To prove that the cross-ratio of the range of points A, B, C, D is equal to the cross-ratio of the pencil of planes* $[lX]$, $[lY]$, $[lZ]$, $[lT]$.

Take $XYZT$ as the tetrahedron of reference, and suppose that the line l meets the planes $x = 0$, $t = 0$ in the points $A(0, y_1, z_1, t_1)$, $D(x_2, y_2, z_2, 0)$ respectively. Then we can put

$$\mathbf{B} \equiv y_2\mathbf{A} - y_1\mathbf{D},$$

$$\mathbf{C} \equiv z_2\mathbf{A} - z_1\mathbf{D},$$

so that (expressing the symbols of the points of the line AD in the parametric form $\theta\mathbf{A} - \mathbf{D}$, for appropriate θ)

$$(D, A, B, C) = (0, \infty, y_2/y_1, z_2/z_1)$$

$$= \frac{y_2 z_1}{y_1 z_2}.$$

Also it is easy to show that the symbol for the *coordinates* of the plane $[lX]$ is

$$\pi_1 \equiv \left(0, \ -z_2, \ y_2, \ \frac{y_1 z_2 - y_2 z_1}{t_1}\right),$$

and that the corresponding symbol for the plane $[lT]$ is

$$\pi_4 \equiv \left(\frac{y_1 z_2 - y_2 z_1}{x_2}, \ z_1, \ -y_1, \ 0\right).$$

Hence the symbols π_2, π_3 of the planes $[lY]$, $[lZ]$ are given in terms of π_1, π_4 by

$$\pi_2 \equiv z_1\pi_1 + z_2\pi_4,$$

$$\pi_3 \equiv y_1\pi_1 + y_2\pi_4.$$

The cross-ratio of the pencil of planes is therefore given by the relation

$$(\pi_4, \pi_1, \pi_2, \pi_3) = (0, \infty, z_1/z_2, y_1/y_2)$$

$$= \frac{y_2 z_1}{y_1 z_2},$$

as required.

Alternative proof. Let the plane $[lY]$ cut XT in P and the plane $[lZ]$ cut XT in Q. Then the cross-ratio of the four planes $[lX]$, $[lY]$, $[lZ]$, $[lT]$ is

$$(X, P, Q, T).$$

This, again, is equal to the cross-ratio of the planes

$$YZX, \quad YZP, \quad YZQ, \quad YZT$$

through the line YZ, and therefore equal to the cross-ratio of the four points in which these planes meet l. In the cases of the planes YZX, YZT these points are D, A respectively. Moreover, the points Y, C, P lie by definition in each of the planes $[lY]$ and XYT, and they are therefore collinear, so that l meets the plane YZP in C; similarly, l meets the plane YZQ in B. The cross-ratio is therefore

$$(D, C, B, A)$$

or

$$(A, B, C, D)$$

on interchanging two pairs of letters. This is the required result.

ILLUSTRATION 1. *Theorem of Desargues. Let ABC, $A'B'C'$ be two (generally situated) triangles in different planes, such that AA', BB', CC' meet in a point T. Then the lines BC, $B'C'$ meet in a point L, the lines CA, $C'A'$ meet in a point M, and the lines AB, $A'B'$ meet in a point N; and L, M, N are collinear.*

First proof. Since BB', CC' meet in T, the points B, B', C, C' are coplanar, and so BC, $B'C'$ do meet, in a point L. Moreover, since L lies on each of the lines BC, $B'C'$, it lies in each of the planes ABC, $A'B'C'$. Similarly, M, N lie in each of these planes. The points L, M, N are therefore collinear, lying in each of the planes ABC, $A'B'C'$.

Second proof. Take A, B, C, T as the vertices X, Y, Z, T of the tetrahedron of reference. Then A', lying on XT, has coordinates $(\alpha, 0, 0, 1)$; similarly B' has coordinates $(0, \beta, 0, 1)$ and C' has coordinates $(0, 0, \gamma, 1)$. The point $(0, \beta, -\gamma, 0)$ lies on each of the lines BC, $B'C'$ and so it is the point L. Similarly, $(-\alpha, 0, \gamma, 0)$ is M and $(\alpha, -\beta, 0, 0)$ is N. The points LMN lie in each of the planes $t = 0$, $x/\alpha + y/\beta + z/\gamma = 0$ and are therefore collinear.

Third proof. Since A' lies on TA, its symbol can be expressed in terms of **T**, **A**, and so (adjusting multiples in **A** if necessary) we may write (cf. § 4, Theorem-example 4, p. 3)

$$\mathbf{A}' \equiv \mathbf{T} + \mathbf{A},$$

and similarly $$\mathbf{B}' \equiv \mathbf{T} + \mathbf{B}, \quad \mathbf{C}' \equiv \mathbf{T} + \mathbf{C}.$$

Hence

$$\mathbf{L} \equiv \mathbf{B} - \mathbf{C} \equiv \mathbf{B}' - \mathbf{C}', \quad \mathbf{M} \equiv \mathbf{C} - \mathbf{A} \equiv \mathbf{C}' - \mathbf{A}', \quad \mathbf{N} \equiv \mathbf{A} - \mathbf{B} \equiv \mathbf{A}' - \mathbf{B}',$$

so that $$\mathbf{L} + \mathbf{M} + \mathbf{N} \equiv \mathbf{0},$$

and the points L, M, N are therefore collinear.

ILLUSTRATION 2. *The polar plane of a point with respect to a tetrahedron. Let $XYZT$ be a tetrahedron and O a point not on any of the faces. Suppose that XO meets the plane YZT in a point X_1, and that points Y_1, Z_1, T_1 are defined similarly. Suppose, finally, that the plane $Y_1Z_1T_1$ meets the plane YZT in a line l_1, and that lines l_2, l_3, l_4 are defined similarly. Then the lines l_1, l_2, l_3, l_4 are coplanar. Their plane is called the* POLAR PLANE *of O with respect to the tetrahedron.*

Take $XYZT$ as the tetrahedron of reference and O as the unit point $(1, 1, 1, 1)$. Then

$$X_1 \equiv (0, 1, 1, 1), \quad Y_1 \equiv (1, 0, 1, 1), \quad Z_1 \equiv (1, 1, 0, 1), \quad T_1 \equiv (1, 1, 1, 0),$$

so that the equation of the plane $Y_1Z_1T_1$ is

$$\begin{vmatrix} x & y & z & t \\ 1 & 0 & 1 & 1 \\ 1 & 1 & 0 & 1 \\ 1 & 1 & 1 & 0 \end{vmatrix} = 0$$

or $$-2x + y + z + t = 0.$$

The equation of any plane through the line l_1 is therefore

$$-2x + y + z + t + \lambda x = 0.$$

When $\lambda = 3$, we obtain the plane

$$x+y+z+t = 0,$$

and, by symmetry, the four lines l_1, l_2, l_3, l_4 lie in it.

Note that this gives us a geometrical construction from the unit point to the unit plane, so that both cannot be chosen arbitrarily.

ILLUSTRATION 3. *Möbius's tetrahedra. $XYZT$, $ABCD$ are two tetrahedra with the property that A, B, C, D lie respectively in the planes YZT, ZXT, XYT, XYZ while X, Y, Z lie respectively in the planes BCD, CAD, ABD. Then T lies in the plane ABC.*

Take $XYZT$ as the tetrahedron of reference and D (as we may) to be the point $(1, 1, 1, 0)$ in the plane XYZ. We may then express the coordinates of A, B, C as follows:

$$\begin{aligned} A &\equiv (\ 0,\ -m_1,\ n_1,\ -1),\\ B &\equiv (\ l_2,\ 0,\ -n_2,\ -1),\\ C &\equiv (-l_3,\ m_3,\ 0,\ -1),\end{aligned}$$

and

$$D \equiv (\ 1,\ 1,\ 1,\ 0).$$

[The minus signs are inserted purely for convenience.]

The plane BCD is given to pass through X. Hence

$$\begin{vmatrix} 1 & 0 & 0 & 0 \\ l_2 & 0 & -n_2 & -1 \\ -l_3 & m_3 & 0 & -1 \\ 1 & 1 & 1 & 0 \end{vmatrix} = 0$$

or $m_3 = n_2$, on reduction. Similarly, $n_1 = l_3$ and $l_2 = m_1$. Hence we can take the coordinates of A, B, C in the form (writing $m_3 = n_2 = p$, etc.)

$$\begin{aligned} A &\equiv (\ 0,\ -r,\ q,\ -1),\\ B &\equiv (\ r,\ 0,\ -p,\ -1),\\ C &\equiv (-q,\ p,\ 0,\ -1).\end{aligned}$$

The symbols of A, B, C, T are thus connected by the syzygy

$$p\mathbf{A}+q\mathbf{B}+r\mathbf{C}+(p+q+r)\mathbf{T} \equiv \mathbf{0},$$

so that T lies in the plane ABC.

Note the 'skew-symmetry' of the coordinates

$$\begin{array}{llll}(0, & -r, & q, & -1), \\ (r, & 0, & -p, & -1), \\ (-q, & p, & 0, & -1), \\ (1, & 1, & 1, & 0).\end{array}$$

It is implicit in the working that the vertices of either tetrahedron are assumed not to lie on the edges of the other; in particular, D is not on YZ, ZX or XY.

ILLUSTRATION 4. *Harmonic inversion.* Let O be a given point and π a given plane not through O. If P is an arbitrary point of space, we can construct a point P' related to it as follows. Let OP meet π in Q; then P' is taken to be the harmonic conjugate of P with respect to O and Q.

The construction from P to P' defines a transformation of P into P' called a *harmonic inversion with respect to the point O and the plane π.*

Choose the tetrahedron of reference such that the vertex T is at O while π is the plane XYZ whose equation is $t = 0$. If P is the point $(\alpha, \beta, \gamma, \delta)$, then Q is the point $(\alpha, \beta, \gamma, 0)$, and so its symbol $\mathbf{Q}$ satisfies the relation

$$\mathbf{P} \equiv \mathbf{Q} + \delta\mathbf{T}.$$

The point P', which is the harmonic conjugate of P with respect to Q and T, is given by the symbol

$$\mathbf{P}' \equiv \mathbf{Q} - \delta\mathbf{T},$$

so that *the coordinates of P' are* $(\alpha, \beta, \gamma, -\delta)$.

Theorem-examples. 1. The point O transforms into itself and each point of the plane π transforms into itself. The point P' transforms into P.

2. A line through O transforms into itself, and corresponding points determine an involution on that line.

[*Note.* Distinguish carefully between a *self-corresponding line* and a *line of self-corresponding points.*]

3. A plane α transforms into a plane α' and a line l transforms into a line l'.

4. Given a tetrahedron of reference $XYZT$ and a point $P(\xi, \eta, \zeta, \tau)$, there are in all four harmonic inverses of P with respect to the vertices X, Y, Z, T and the respective planes $x = 0$, $y = 0$, $z = 0$, $t = 0$; the inverse points are

$$P_1 \equiv (-\xi, \eta, \zeta, \tau), \quad P_2 \equiv (\xi, -\eta, \zeta, \tau), \quad P_3 \equiv (\xi, \eta, -\zeta, \tau), \quad P_4 \equiv (\xi, \eta, \zeta, -\tau).$$

Consider next another construction from an arbitrary point P of space to a point P', as follows. Let l, m be two given lines, and let the unique transversal from P meet them in L, M respectively; then P' is taken to be the harmonic conjugate of P with respect to L and M.

This transformation of P into P' is called a *harmonic inversion with respect to the lines l and m.*

Take a tetrahedron of reference such that l is the line XT and m the line YZ. If P is the point $(\alpha, \beta, \gamma, \delta)$, then L is the point $(\alpha, 0, 0, \delta)$ and M is the point $(0, \beta, \gamma, 0)$. The symbols of P, L, M are connected by the syzygy

$$\mathbf{P} \equiv \mathbf{L} + \mathbf{M},$$

and so the point P', which is the harmonic conjugate of P with respect to L and M, is given by the symbol

$$\mathbf{P}' \equiv \mathbf{L} - \mathbf{M}.$$

The coordinates of P' are therefore $(\alpha, -\beta, -\gamma, \delta)$.

Theorem-examples. 1. Each of the lines l, m is a line of self-corresponding points. The point P' transforms into P.

2. A line meeting l and m transforms into itself, and corresponding points determine an involution on it.

3. A plane α transforms into a plane α', and a line p transforms into a line p'.

4. Given a tetrahedron of reference $XYZT$ and a point $P(\xi, \eta, \zeta, \tau)$, there are in all three harmonic inverses of P with respect to the edges XT, YZ; YT, ZX; ZT, XY respectively, and the inverse points are

$$P_{23} \equiv (\xi, -\eta, -\zeta, \tau), \quad P_{31} \equiv (-\xi, \eta, -\zeta, \tau), \quad P_{12} \equiv (-\xi, -\eta, \zeta, \tau).$$

5. If any one of the eight points $P, P_1, P_2, P_3, P_4, P_{23}, P_{31}, P_{12}$ is inverted with respect to a vertex and a face of the tetrahedron of reference, or with respect to two opposite edges, the inverse point is also one of the eight points.

ILLUSTRATION 5. *Desmic tetrahedra. Two tetrahedra so related that each edge of either meets two opposite edges of the other are said to be* DESMIC.

We establish the existence of such a configuration and some of its simpler properties.

Let $XYZT$ be a tetrahedron and D a point not on any of its faces. Suppose that the syzygy connecting the symbols $\mathbf{X}$, $\mathbf{Y}$, $\mathbf{Z}$, $\mathbf{T}$, $\mathbf{D}$ is

$$\mathbf{D} \equiv \mathbf{X}+\mathbf{Y}+\mathbf{Z}+\mathbf{T}.$$

If A, B, C are the harmonic inverses of D with respect to the lines XT, YZ; YT, ZX; ZT, XY respectively, then

$$\mathbf{A} \equiv \mathbf{X}-\mathbf{Y}-\mathbf{Z}+\mathbf{T},$$
$$\mathbf{B} \equiv -\mathbf{X}+\mathbf{Y}-\mathbf{Z}+\mathbf{T},$$
$$\mathbf{C} \equiv -\mathbf{X}-\mathbf{Y}+\mathbf{Z}+\mathbf{T}.$$

We prove that $XYZT$, $ABCD$ are desmic tetrahedra. By definition of the harmonic inverses, each of the lines AD, BD, CD meets two edges of $XYZT$. Moreover, the syzygies

$$\mathbf{B}+\mathbf{C} \equiv -2\mathbf{X}+2\mathbf{T},$$
$$\mathbf{B}-\mathbf{C} \equiv 2\mathbf{Y}-2\mathbf{Z}$$

show that BC meets both XT and YZ, with similar results for CA, AB. Hence each of AD, BC meets each of XT, YZ; each of BD, CA meets each of YT, ZX; each of CD, AB meets each of ZT, XY. The two tetrahedra are therefore desmic.

We next prove that *the two tetrahedra $XYZT$, $ABCD$ are in fourfold perspective.*

(i) The syzygies

$$\mathbf{A}-2\mathbf{X} \equiv \mathbf{B}-2\mathbf{Y} \equiv \mathbf{C}-2\mathbf{Z} \equiv -\mathbf{D}+2\mathbf{T} \equiv -\mathbf{X}-\mathbf{Y}-\mathbf{Z}+\mathbf{T}$$

show that AX, BY, CZ, DT pass through the point S whose symbol is given by

$$\mathbf{S} \equiv \mathbf{X}+\mathbf{Y}+\mathbf{Z}-\mathbf{T}.$$

(ii) The syzygies

$$-\mathbf{A}+2\mathbf{T} \equiv \mathbf{B}+2\mathbf{Z} \equiv \mathbf{C}+2\mathbf{Y} \equiv \mathbf{D}-2\mathbf{X} \equiv -\mathbf{X}+\mathbf{Y}+\mathbf{Z}+\mathbf{T}$$

show that AT, BZ, CY, DX pass through the point P whose symbol is given by

$$\mathbf{P} \equiv -\mathbf{X}+\mathbf{Y}+\mathbf{Z}+\mathbf{T}.$$

(iii) The syzygies

$$\mathbf{A}+2\mathbf{Z} \equiv -\mathbf{B}+2\mathbf{T} \equiv \mathbf{C}+2\mathbf{X} \equiv \mathbf{D}-2\mathbf{Y} \equiv \mathbf{X}-\mathbf{Y}+\mathbf{Z}+\mathbf{T}$$

show that AZ, BT, CX, DY pass through the point Q whose symbol is given by

$$\mathbf{Q} \equiv \mathbf{X}-\mathbf{Y}+\mathbf{Z}+\mathbf{T}.$$

(iv) The syzygies

$$\mathbf{A}+2\mathbf{Y}\equiv\mathbf{B}+2\mathbf{X}\equiv-\mathbf{C}+2\mathbf{T}\equiv\mathbf{D}-2\mathbf{Z}\equiv\mathbf{X}+\mathbf{Y}-\mathbf{Z}+\mathbf{T}$$

show that AY, BX, CT, DZ pass through the point R whose symbol is given by

$$\mathbf{R}\equiv\mathbf{X}+\mathbf{Y}-\mathbf{Z}+\mathbf{T}.$$

It is now a simple matter to prove that any two of the tetrahedra $XYZT$, $ABCD$, $PQRS$ are in perspective (for appropriate orders of the vertices) from each vertex of the third. The system of three tetrahedra is completely symmetrical; *each of its three pairs consists of two desmic tetrahedra.*

Moreover, we may prove that *the twelve points of intersection of the sides of these tetrahedra also form, when suitably grouped, another system of three desmically related tetrahedra.*

For the syzygies

$$\mathbf{B}+\mathbf{C}\equiv-2\mathbf{X}+2\mathbf{T},\quad \mathbf{B}-\mathbf{C}\equiv\ \ 2\mathbf{Y}-2\mathbf{Z}$$

show that BC meets XT and YZ in points which we may conveniently take as $\mathbf{X}-\mathbf{T}$ and $\mathbf{Y}-\mathbf{Z}$; and the syzygies

$$\mathbf{A}+\mathbf{D}\equiv\ \ 2\mathbf{X}+2\mathbf{T},\quad \mathbf{A}-\mathbf{D}\equiv-2\mathbf{Y}-2\mathbf{Z}$$

provide similarly the points $\mathbf{X}+\mathbf{T}$ and $\mathbf{Y}+\mathbf{Z}$. Permuting cyclically, we obtain twelve points in all, which we name according to the scheme

$$\begin{array}{lll}
\mathbf{X}'\equiv\mathbf{X}-\mathbf{T}, & \mathbf{A}'\equiv\mathbf{Y}-\mathbf{T}, & \mathbf{P}'\equiv\mathbf{Z}-\mathbf{T},\\
\mathbf{Y}'\equiv\mathbf{X}+\mathbf{T}, & \mathbf{B}'\equiv\mathbf{Y}+\mathbf{T}, & \mathbf{Q}'\equiv\mathbf{Z}+\mathbf{T},\\
\mathbf{Z}'\equiv\mathbf{Y}+\mathbf{Z}, & \mathbf{C}'\equiv\mathbf{Z}+\mathbf{X}, & \mathbf{R}'\equiv\mathbf{X}+\mathbf{Y},\\
\mathbf{T}'\equiv\mathbf{Y}-\mathbf{Z}, & \mathbf{D}'\equiv\mathbf{Z}-\mathbf{X}, & \mathbf{S}'\equiv\mathbf{X}-\mathbf{Y}.
\end{array}$$

The three tetrahedra $X'Y'Z'T'$, $A'B'C'D'$, $P'Q'R'S'$ may easily be proved to be perspectively related in the same way as $XYZT$, $ABCD$, $PQRS$. They form a desmic system *associated* with the first.

EXAMPLES I

1. Prove that the points

$$A(1, 2, 3, 4),\quad B(4, 3, 2, 1),\quad C(1, 1, 1, 1),\quad D(3, 1, -1, -3)$$

are collinear and that A, B separate C, D harmonically. Prove also that the line meets the plane $t=0$ in the point $(3, 2, 1, 0)$.

2. Prove that the plane through the points $(1, 2, 2, 1)$, $(0, 1, 2, 3)$, $(1, 3, 3, 1)$ contains the points $(2, 6, 7, 5)$ and $(0, 3, 4, 3)$, and that it meets the line joining the points $(4, 2, 0, 1)$, $(4, 4, 3, 4)$ in the point $(0, 2, 3, 3)$.

3. Prove that the equations of the four planes which pass through the line $2x+3y-4z-t = 0$, $x-y+z+5t = 0$ and contain a vertex of the tetrahedron of reference are

$$5y-6z-11t = 0, \quad 5x-z+14t = 0, \quad 6x-y+19t = 0, \quad 11x+14y-19z = 0.$$

4. Prove that the four planes

$$x+y-z = 0, \quad x-y-z+t = 0, \quad y+2z-2t = 0, \quad x+5z-4t = 0$$

have a common point.

What can you say about the four points

$$(1, 1, -1, 0), \quad (1, -1, -1, 1), \quad (0, 1, 2, -2), \quad (1, 0, 5, -4)?$$

5. Prove that the equation of the point in which the line

$$a_1x+b_1y+c_1z+d_1t = 0, \quad a_2x+b_2y+c_2z+d_2t = 0$$

meets the plane $$px+qy+rz+st = 0$$

is
$$\begin{vmatrix} l & m & n & p \\ a_1 & b_1 & c_1 & d_1 \\ a_2 & b_2 & c_2 & d_2 \\ p & q & r & s \end{vmatrix} = 0.$$

6. Prove that the transformation from a coordinate system with tetrahedron of reference X, Y, Z, T to a system with tetrahedron

$$X', \quad Y', \quad Z', \quad T',$$

in which the planes $x' = 0$, $y' = 0$, $z' = 0$, $t' = 0$ are the same as the planes $x = 0$, $x+y = 0$, $x+y+z = 0$, $x+y+z+t = 0$, while the same point has coordinates $(1, 1, 1, 1)$, in each system, is given by the relations

$$x = x', \quad y = 2y'-x', \quad z = 3z'-2y', \quad t = 4t'-3z'.$$

7. Prove that the line joining the points $(\theta^3, \theta^2, \theta, 1)$, $(\phi^3, \phi^2, \phi, 1)$ lies entirely in each of the planes

$$x-y(\theta+\phi)+z\theta\phi = 0, \quad y-z(\theta+\phi)+t\theta\phi = 0.$$

8. Prove that, if θ and ϕ are connected by the relation

$$a\theta\phi+b(\theta+\phi)+c = 0,$$

then the coordinates of any point on the line joining the points

$$(\theta^3, \theta^2, \theta, 1), \quad (\phi^3, \phi^2, \phi, 1)$$

satisfy the equation

$$a(y^2-zx)+b(yz-xt)+c(z^2-yt) = 0.$$

9. Prove that the four points in which the line joining the points $(1, 2, 4, -6)$, $(1, 8, 7, 3)$ meets the faces of the tetrahedron of reference from a harmonic range.

10. Prove that, if α and β are unequal, the two lines

$$x - 2y\alpha + z\alpha^2 = 0, \quad y - 2z\alpha + t\alpha^2 = 0$$

and

$$x - 2y\beta + z\beta^2 = 0, \quad y - 2z\beta + t\beta^2 = 0$$

are skew.

11. Prove that the three tetrahedra whose faces are

$$x \pm t = 0, \quad y \pm z = 0;$$

$$y \pm t = 0, \quad z \pm x = 0;$$

$$z \pm t = 0, \quad x \pm y = 0$$

form a desmic system.

MISCELLANEOUS EXAMPLES I

1. Two tetrahedra $XYZT$, $PQRS$ are such that each edge of one meets two opposite edges of the other, i.e. PQ and RS meet XY and ZT, QR and PS meet YZ and XT, RP and QS meet ZX and YT. Show that any face of one cuts the other tetrahedron in a complete quadrilateral whose diagonals are the edges in that face, and hence that any edge of one is divided harmonically by the points in which it meets the two edges of the other.
Show also that PT, QZ, RY, SX are concurrent. [M.T. I.]

2. Three non-intersecting lines are given in space. Show that through any point on any one of these lines one and only one line may be drawn intersecting the other two lines, and that no two of the lines so drawn meet.
$A_1A_2A_3$, $B_1B_2B_3$ and $C_1C_2C_3$ are three non-intersecting straight lines; $A_1B_1C_1$, $A_2B_2C_2$ and $A_3B_3C_3$ are three transversals. Show that the lines A_3B_2 and B_1C_2 do not intersect. [M.T. I.]

3. A, B, C, D, E are five points in space of which no four are coplanar. From each of them three lines are drawn which meet the pairs of opposite edges of the tetrahedron formed from the four remaining points. Show that the three lines of this set (not including the edges of the tetrahedra), which meet AB, all do so in the same point and are coplanar. [M.T. I.]

4. The line joining a point P to the vertex A of a tetrahedron $ABCD$ meets the face BCD in H, and Q is the harmonic conjugate of P with respect to A and H. The line BQ meets the face CDA in K, and R is the harmonic conjugate of Q with respect to B and K. Prove that the line PR meets the edges AB, CD and that the points of intersection are harmonic conjugates with respect to P and R. [M.T. I.]

5. Two triangles $ABC, A'B'C'$ are in perspective in space of three dimensions from a point O. Prove that the three points $(BC, B'C')$, $(CA, C'A')$, $(AB, A'B')$ are collinear (on the *axis of perspective*).

Prove that the four axes for the pairs of triangles

$$ABC, A'B'C'; \quad A'BC, AB'C'; \quad AB'C, A'BC'; \quad ABC', A'B'C$$

are coplanar.

From nine points lying in sets of three on concurrent lines are formed thirty-six sets of three perspective triangles. Show that the three axes of the pairs of such a set of three distinct triangles meet in a point, and that the thirty-six points thus obtained from the thirty-six sets of triangles lie in sets of four on twenty-seven lines. [P.]

6. Five skew lines $l_1, l_2, \ldots, l_5$ are given in general position in space of three dimensions. The transversals from a variable point P_1 on l_1 to l_2, l_3 and to l_4, l_5 meet these lines respectively in P_2, P_3 and P_4, P_5. Show that the locus of the point of intersection of P_2P_4 and P_3P_5 is a straight line. [P.]

7. $ABCD$ is a tetrahedron, and S is a point not lying in any of its faces. The transversals from S to the pairs of opposite edges BC, AD; CA, BD; AB, CD meet these edges respectively in P, P'; Q, Q'; R, R'. Prove that AP, BQ, CR, DS are concurrent. [M.T. I.]

APPENDIX TO CHAPTER I

A configuration in space which relates the lines and circles of a triangle to the vertices of a desmic system of tetrahedra.

1. A somewhat unexpected use of homogeneous coordinates in space is found by applying the ideas of the text to the circles of an ordinary Euclidean plane. The author has discussed this application elsewhere,* but the following account is self-contained; we do, however, assume familiarity with the pure geometry of the triangle and with Cartesian analytical geometry in a plane. There is little doubt that anyone reading this book will have the necessary knowledge.

If we write $$S \equiv a(x^2+y^2)+2px+2qy+r,$$

then the equation $S = 0$ represents (in a Cartesian plane) a circle, which may be real or 'imaginary'; it may also, in special cases, be a straight line or a point. It is convenient to denote the circle itself by the same letter S.

* *Mathematical Gazette*, 21 (1937), pp. 46–9. See also an article by Dr D. Pedoe in the same volume, pp. 210–15.

Now regard the numbers a, p, q, r as the homogeneous coordinates of a point (a, p, q, r) in space of three dimensions. We use the notation $\mathbf{S}$ (in bold type) to denote both the point and also the symbol which defines it.

If S, S' are two circles, then the equation

$$S+\lambda S' = 0$$

determines a *pencil* of circles, or, as it is called, a *coaxal system*. In the particular case when S, S' are *both* straight lines (the coefficient of x^2 and y^2 vanishing in each equation) the equation determines a pencil of straight lines in the ordinary sense of the phrase. The points corresponding to the circles or straight lines of a pencil lie, in the three-dimensional space, on a straight line,* which we can describe as *representing* the pencil. The straight lines which represent two pencils are skew unless those pencils have a circle or straight line in common; if the first pencil consists of circles through two points A, B and the second of circles through two points A', B', then the case of exception arises when A, B, A', B' are concyclic (or collinear).

2. Let ABC be a triangle which, for convenience of language, we take to be acute-angled and scalene. We denote the sides by the letters a, b, c; the altitudes AP, BQ, CR, which meet in the orthocentre H, by the letters p, q, r; and the sides QR, RP, PQ of the pedal triangle by the letters p', q', r'. We also name certain circles, which occur frequently, as follows:

$$HBC \equiv u, \qquad HCA \equiv v, \qquad HAB \equiv w;$$

$$ABC \equiv t, \text{ the circumcircle;}$$

$$PQR \equiv \delta, \text{ the nine-points circle;}$$

$$AQHR \equiv \alpha, \quad BRHP \equiv \beta, \quad CPHQ \equiv \gamma;$$

$$BCQR \equiv \alpha', \quad CARP \equiv \beta', \quad ABPQ \equiv \gamma'.$$

The points in three dimensions which represent these circles are denoted by the same letters in bold type.

* The coordinates of a typical point are

$$(a+\lambda a', \quad p+\lambda p', \quad q+\lambda q', \quad r+\lambda r').$$

3. We require a basic system of four circles in terms of whose symbols those of the others can be expressed. We choose the circles u, v, w, t and, for brevity in exposition, we assume that the coefficients of x^2 and y^2 in their four equations are taken as unity—they are not zero, so the assumption is legitimate. We note also that the four points $\mathbf{u}$, $\mathbf{v}$, $\mathbf{w}$, $\mathbf{t}$ are not coplanar; if they were, the pencils defined by v, w and by u, t would have a common circle, so that H, A, B, C would be concyclic, which is not so.

4. Our first task is to express the symbols of the points corresponding to the circles enumerated above in terms of $\mathbf{u}$, $\mathbf{v}$, $\mathbf{w}$, $\mathbf{t}$. Since p is on the radical axis of v, w, $\mathbf{p}$ is a multiple of $\mathbf{v}-\mathbf{w}$ and likewise $\mathbf{a}$ of $\mathbf{u}-\mathbf{t}$, so we can write

$$\mathbf{p}\equiv\mathbf{v}-\mathbf{w}, \quad \mathbf{q}\equiv\mathbf{w}-\mathbf{u}, \quad \mathbf{r}\equiv\mathbf{u}-\mathbf{v}$$

$$\mathbf{a}\equiv\mathbf{u}-\mathbf{t}, \quad \mathbf{b}\equiv\mathbf{v}-\mathbf{t}, \quad \mathbf{c}\equiv\mathbf{w}-\mathbf{t}.$$

We pause to remark that these symbols define six points which are the vertices of the quadrilateral formed by four lines pqr, pbc, qca, rab.

Consider next the circles α, β, γ. The circle β, through H, B, is given by a symbol of the form $\mathbf{u}+\lambda\mathbf{w}$, and the circle γ, through H, C, is given by a symbol of the form $\mathbf{u}+\lambda'\mathbf{v}$. But the circles β, γ have radical axis p whose symbol is $\mathbf{v}-\mathbf{w}$, and so $\lambda'=\lambda$. Proceeding cyclically, we find relations of the form

$$\boldsymbol{\beta}\equiv\mathbf{u}+\lambda\mathbf{w}, \quad \boldsymbol{\gamma}\equiv\mathbf{u}+\lambda\mathbf{v},$$

$$\boldsymbol{\gamma}\equiv\mathbf{v}+\mu\mathbf{u}, \quad \boldsymbol{\alpha}\equiv\mathbf{v}+\mu\mathbf{w},$$

$$\boldsymbol{\alpha}\equiv\mathbf{w}+\nu\mathbf{v}, \quad \boldsymbol{\beta}\equiv\mathbf{w}+\nu\mathbf{u},$$

so that $\lambda=\mu=\nu=1$. Hence

$$\boldsymbol{\alpha}\equiv\mathbf{v}+\mathbf{w}, \quad \boldsymbol{\beta}\equiv\mathbf{w}+\mathbf{u}, \quad \boldsymbol{\gamma}\equiv\mathbf{u}+\mathbf{v}.$$

Similarly, we may prove that

$$\boldsymbol{\alpha}'\equiv\mathbf{u}+\mathbf{t}, \quad \boldsymbol{\beta}'\equiv\mathbf{v}+\mathbf{t}, \quad \boldsymbol{\gamma}'\equiv\mathbf{w}+\mathbf{t}.$$

We now evaluate the symbols $\mathbf{p}'$, $\mathbf{q}'$, $\mathbf{r}'$. The line p' belongs to the pencil defined by b, q and also to the pencil defined by c, r. It is therefore easy to obtain relations which we may take in the form

$$\mathbf{p}'\equiv\mathbf{u}-\mathbf{v}-\mathbf{w}+\mathbf{t}, \quad \mathbf{q}'\equiv-\mathbf{u}+\mathbf{v}-\mathbf{w}+\mathbf{t}, \quad \mathbf{r}'\equiv-\mathbf{u}-\mathbf{v}+\mathbf{w}+\mathbf{t}.$$

Finally, the nine-points circle passes through Q, R and so belongs to the pencil defined by α, α'. Hence, using symmetry, we have

$$\boldsymbol{\delta} \equiv \mathbf{u}+\mathbf{v}+\mathbf{w}+\mathbf{t}.$$

The point $\boldsymbol{\delta}$ in the three-dimensional space is therefore the 'unit' point for our coordinate system.

5. We are now in a position to describe some of the features of the configuration in space. We can regard it as defined by the four points $\mathbf{u}$, $\mathbf{v}$, $\mathbf{w}$, $\mathbf{t}$, together with the unit point $\boldsymbol{\delta}$. The points $\boldsymbol{\alpha}$, $\boldsymbol{\alpha}'$ are where the transversal from $\boldsymbol{\delta}$ to the opposite edges **vw**, **ut** of the tetrahedron **uvwt** meets those edges.

We note in passing the subsidiary result that *the triangles* $\boldsymbol{\alpha\beta\gamma}$, $\boldsymbol{\alpha'\beta'\gamma'}$, *which represent the sets of circles AQHR, BRHP, CPHQ and BCQR, CARP, ABPQ respectively, are in perspective from the point* $\boldsymbol{\delta}$ *which represents the nine-points circle.* The axis of perspective is the straight line on which lie the points representing the altitudes of the triangle ABC, since we have the identity

$$-(\boldsymbol{\beta}-\boldsymbol{\gamma}) \equiv (\boldsymbol{\beta}'-\boldsymbol{\gamma}') \equiv \mathbf{p}.$$

Examination of the symbols shows at once that **uvwt** and $\mathbf{p'q'r'}\boldsymbol{\delta}$ are *desmic* tetrahedra, each edge of the one meeting a pair of opposite edges of the other. Taking symmetrically typical edges, the intersections are indicated by the scheme

	vw	**ut**
$\mathbf{q'r'}$	$\mathbf{p}$	$\mathbf{a}$
$\boldsymbol{\delta}\mathbf{p'}$	$\boldsymbol{\alpha}$	$\boldsymbol{\alpha}'$

6. We proved in the text that, when two desmic tetrahedra **uvwt** and $\mathbf{p'q'r'}\boldsymbol{\delta}$ are given, there also exists a third tetrahedron, say $\boldsymbol{\lambda\mu\nu\rho}$, such that the pairs selected from the three tetrahedra are all desmic; any two of these tetrahedra are in perspective from each vertex of the third. The vertices of this remaining tetrahedron are, in fact, given by the relations (compare pp. 21–2)

$$\begin{aligned} \boldsymbol{\lambda} &\equiv -\mathbf{u}+\mathbf{v}+\mathbf{w}+\mathbf{t}, \\ \boldsymbol{\mu} &\equiv \mathbf{u}-\mathbf{v}+\mathbf{w}+\mathbf{t}, \\ \boldsymbol{\nu} &\equiv \mathbf{u}+\mathbf{v}-\mathbf{w}+\mathbf{t}, \\ \boldsymbol{\rho} &\equiv \mathbf{u}+\mathbf{v}+\mathbf{w}-\mathbf{t}, \end{aligned}$$

and we have to find the circles which they represent.

Consider first the circle λ. Since

$$\boldsymbol{\lambda} \equiv \boldsymbol{\alpha} - \mathbf{a} \equiv 2\mathbf{t} - \mathbf{p}',$$

λ passes through the points of intersection of α, a (the fact that they are 'imaginary' is irrelevant to the argument) and of t, p'. But the circle whose centre is A and the square of whose radius is

$$AB.AR = AH.AP = AC.AQ$$

inverts α into a and t into p'; it therefore passes through the four points just described and so it is the required circle λ. Hence *λ is the circle whose centre is A and the square of whose radius is the power of A with respect to each of the circles α', β, γ.* The circles μ and ν can be defined similarly.

We now determine the circle ρ; unfortunately, it is 'imaginary', but that imposes no essential limitation on the configuration.* We can apply reasoning similar to that just given, noting (i) the identities

$$\boldsymbol{\rho} \equiv \boldsymbol{\alpha} + \mathbf{a} \equiv 2\mathbf{u} - \mathbf{p}',$$

and (ii) that the circle whose centre is H and the square of whose radius is *minus* $HA.HP = HB.HQ = HC.HR$ inverts α into a and u into p'. Hence *ρ is the circle whose centre is H and the square of whose radius is the power of H with respect to each of the circles α', β', γ'.*

7. Having obtained these three tetrahedra of a desmic system, we proceed to seek the three further tetrahedra, mutually related in exactly the same way, which form the *associated desmic system.* We have introduced twenty-four circles in all; twelve of them account for the first desmic system, and the others form the associated system. The three new tetrahedra are $\mathbf{ap}\boldsymbol{\alpha\alpha}'$, $\mathbf{bq}\boldsymbol{\beta\beta}'$, $\mathbf{cr}\boldsymbol{\gamma\gamma}'$ respectively.

The proof of this statement follows so directly from the symbols that we leave it as an exercise for the reader.

8. It is of interest to consider the result of inverting the whole plane configuration with respect to the circles λ, μ, ν, ρ and to study

* It is perhaps surprising that, of all the circles considered so far, only this one (in the case of an acute-angled triangle ABC) is 'imaginary'.

the corresponding process in space. It will be sufficient to take the circle ρ for illustration; λ, μ, ν are essentially equivalent, but the use of ρ gives us greater symmetry of notation.

We have the following inverse pairs:

$$(u,p'),\quad (v,q'),\quad (w,r'),\quad (t,\delta),\quad (a,\alpha),\quad (b,\beta),\quad (c,\gamma).$$

In addition, the circles $p, q, r, \alpha', \beta', \gamma', \lambda, \mu, \nu$ invert into themselves, as is easy to verify, and the circle ρ inverts into itself point for point.

To describe the corresponding transformation in space, we suppose that the symbol of a typical point is expressed in the form

$$\xi\mathbf{u}+\eta\mathbf{v}+\zeta\mathbf{w}+\tau\mathbf{t},$$

and we call it 'the point (ξ, η, ζ, τ)'. For example, the points **a** and $\mathbf{p}'$ are $(1, 0, 0, -1)$ and $(1, -1, -1, 1)$ respectively. We can prove that, if the points (ξ, η, ζ, τ), $(\xi', \eta', \zeta', \tau')$ represent circles inverse with respect to ρ, then these two points correspond in the transformation

$$\begin{aligned} k\xi' &= \xi-\eta-\zeta+\tau, \\ k\eta' &= -\xi+\eta-\zeta+\tau, \\ k\zeta' &= -\xi-\eta+\zeta+\tau, \\ k\tau' &= \xi+\eta+\zeta+\tau, \end{aligned}$$

where k is a coefficient of proportionality. The self-corresponding points in this correspondence are (i) the point $(1, 1, 1, -1)$ which gives $\boldsymbol{\rho}$ itself, and (ii) all the points of the plane whose equation is

$$\xi+\eta+\zeta-\tau = 0.$$

This plane contains the nine points $\mathbf{p}$, $\mathbf{q}$, $\mathbf{r}$, $\boldsymbol{\alpha}'$, $\boldsymbol{\beta}'$, $\boldsymbol{\gamma}'$, $\boldsymbol{\lambda}$, $\boldsymbol{\mu}$, $\boldsymbol{\nu}$ corresponding to the unaltered circles enumerated above. They form the six vertices of the quadrilateral whose sides are the lines **pqr**, $\mathbf{p}\boldsymbol{\beta}'\boldsymbol{\gamma}'$, $\mathbf{q}\boldsymbol{\gamma}'\boldsymbol{\alpha}'$, $\mathbf{r}\boldsymbol{\alpha}'\boldsymbol{\beta}'$ and the three vertices $\boldsymbol{\lambda}$, $\boldsymbol{\mu}$, $\boldsymbol{\nu}$ of its diagonal triangle. The transformation itself is a *harmonic inversion* in which $\boldsymbol{\rho}$ is the fundamental point and the plane $\xi+\eta+\zeta-\tau = 0$ the fundamental plane. Thus *inversion with respect to the circle* ρ *is represented by harmonic inversion with respect to the point* $\boldsymbol{\rho}$ *and the plane* $\boldsymbol{\lambda\mu\nu}$.

It may be remarked that the three tetrahedra $\mathbf{ap}\boldsymbol{\alpha\alpha}'$, $\mathbf{bq}\boldsymbol{\beta\beta}'$, $\mathbf{cr}\boldsymbol{\gamma\gamma}'$ of the associated desmic system are all unchanged by this transformation. For example, in $\mathbf{ap}\boldsymbol{\alpha\alpha}'$ the vertices $\mathbf{a}$ and $\boldsymbol{\alpha}$ are interchanged, while $\mathbf{p}$ and $\boldsymbol{\alpha}'$ are unaltered.

9. The machinery set up in this way enables us to obtain a number of results in Euclidean geometry, of which one example may be given in illustration:

The radical axis of the circumcircle t and the nine-points circle δ is given by the symbol $\mathbf{u}+\mathbf{v}+\mathbf{w}-3\mathbf{t}$, which can be expressed in the alternative forms $2\mathbf{a}-\mathbf{p}'$, $2\mathbf{b}-\mathbf{q}'$, $2\mathbf{c}-\mathbf{r}'$. Hence *the radical axis of the circumcircle and the nine-points circle of a triangle is the polar line of the orthocentre with respect to the triangle.*

Further suggestions occur in another article by the author in the *Mathematical Gazette*, 31 (1947), pp. 266–9.

CHAPTER II

THE QUADRIC SURFACE

1. Introduction. The points whose coordinates x, y, z, t satisfy a single homogeneous equation

$$f(x, y, z, t) = 0$$

are said to lie on a *surface*, and the degree of that equation is called the *order* of the surface. In particular, the quadratic equation

$$ax^2 + by^2 + cz^2 + dt^2 + 2fyz + 2gzx + 2hxy + 2uxt + 2vyt + 2wzt = 0$$

defines a *quadric surface*. Before proceeding to the fundamental geometrical properties which give to this surface its importance and its richness in results, we prove certain algebraic theorems which we shall often require.

2. The expression of a quadratic form as a 'sum of squares'. The work which we are about to give can be found in most text-books on linear algebra, but it may be convenient to repeat it here.

The result which we prove is the following:

If Q is a quadratic function of n variables $x, y, z, \ldots$, then Q can be expressed in the form

$$Q \equiv a_1 L_1^2 + a_2 L_2^2 + \ldots + a_n L_n^2,$$

where $L_1, L_2, \ldots, L_n$ are linearly independent linear forms in $x, y, z, \ldots$ and $a_1, a_2, \ldots, a_n$ are constants, some of which may, in special cases, be zero.

We deal in succession with the cases $n = 2, 3, 4$ in which we are most interested.

(i) *The case $n = 2$.* The quadratic form is

$$Q \equiv ax^2 + 2hxy + by^2.$$

If a, b are not both zero, we may suppose $a \neq 0$. Then

$$aQ = (ax + hy)^2 + (ab - h^2) y^2,$$

and the result follows on dividing by a.

If $a = b = 0$, we make the transformation

$$x = x', \quad y = lx' + my' \quad (l \neq 0).$$

Then $$Q = 2hlx'^2 + 2hmx'y' \quad (h \neq 0),$$

which can be expressed as the sum of the squares of two functions which are linear in x', y' and therefore in x, y.

Note. Obviously m also cannot vanish, since the determinant of the transformation must not be zero.

(ii) *The case* $n = 3$. The quadratic form is

$$Q \equiv ax^2 + by^2 + cz^2 + 2fyz + 2gzx + 2hxy.$$

If a, b, c are not all zero, we may suppose $a \neq 0$. Then

$$aQ = (ax + hy + gz)^2 + (ab - h^2)\,y^2 + 2(af - gh)\,yz + (ca - g^2)\,z^2,$$

and the result follows on dividing by a and using the theorem for $n = 2$.

If $a = b = c = 0$, we may suppose that f, g, h do not all vanish, so that, say, $g \neq 0$. We make the transformation*

$$x = x', \quad y = y', \quad z = lx' + my' + nz' \quad (l \neq 0).$$

Then $$Q = 2(fy' + gx')(lx' + my' + nz') + 2hx'y'$$
$$= 2glx'^2 + \ldots,$$

which can be expressed as the sum of the squares of three functions which are linear in x', y', z' and therefore in x, y, z.

Note. The determinant of the coefficients in the transformation must not vanish, and so $n \neq 0$.

(iii) *The case* $n = 4$. The quadratic form is

$$Q \equiv ax^2 + by^2 + cz^2 + dt^2 + 2fyz + 2gzx + 2hxy + 2uxt + 2vyt + 2wzt.$$

If a, b, c, d are not all zero, we may suppose $a \neq 0$. Then

$$aQ = (ax + hy + gz + ut)^2 + R,$$

where R is a quadratic function of the three variables y, z, t, and the result follows on dividing by a and using the theorem for $n = 3$ applied to the function R.

* The double use of the letter n will not confuse.

If $a = b = c = d = 0$, we may suppose that f, g, h, u, v, w do not all vanish, so that one of them, say u, is not zero. We then make the transformation

$$x = x', \quad y = y', \quad z = z', \quad t = lx' + my' + nz' + pt' \quad (l \neq 0).$$

Then

$$Q = 2fy'z' + 2gz'x' + 2hx'y' + 2(ux' + vy' + wz')(lx' + my' + nz' + pt')$$
$$= 2ulx'^2 + \ldots,$$

which can be expressed as the sum of the squares of four functions which are linear in x', y', z', t' and therefore in x, y, z, t.

Note. As before, p must not vanish.

We add that the independence of the linear forms in the general case is established by quoting the particular examples

$$n = 2, \quad ax^2 + by^2,$$
$$n = 3, \quad ax^2 + by^2 + cz^2,$$
$$n = 4, \quad ax^2 + by^2 + cz^2 + dt^2,$$

so that the full number of linearly independent forms is certainly required, though in special cases (e.g. $a = 0$ in the above) fewer may suffice. The detailed proof is, however, not quite obvious. The reader should consult *Algebra*, by W. L. Ferrar (Oxford University Press, 1941), p. 146.

A similar result will be obtained from a geometrical point of view in § 17 (p. 47).

3. A simple form for the equation of a quadric. It follows at once from the theorem of § 2 that a transformation can be found which will reduce the equation of a quadric to the standard form

$$ax^2 + by^2 + cz^2 + dt^2 = 0.$$

In the general case, a, b, c, d do not vanish and then we can choose the unit point in such a way as to give the very simple form

$$x^2 + y^2 + z^2 + t^2 = 0.$$

We shall have more to say about these equations later.

In order to be clear about the exceptions which may arise, we consider briefly the cases when one or more of the coefficients vanish.

(i) *The case* $a = b = c = 0$. The equation is

$$t^2 = 0,$$

which represents a *degenerate quadric* consisting of the plane $t = 0$ 'counted twice'.

(ii) *The case* $b = c = 0$. The equation is

$$ax^2 + dt^2 = 0$$

or

$$(x\sqrt{a} + it\sqrt{d})(x\sqrt{a} - it\sqrt{d}) = 0, \qquad (i^2 = -1)$$

which represents a degenerate quadric consisting of the two distinct planes $x\sqrt{a} \pm it\sqrt{d} = 0$. Their line of intersection is the line $x = 0$, $t = 0$.

(iii) *The case* $d = 0$. The equation is

$$ax^2 + by^2 + cz^2 = 0,$$

and this represents a surface with some characteristic features which we must now consider.

We observe first that, if the point $P(\xi, \eta, \zeta, \tau)$ is on the surface, then

$$a\xi^2 + b\eta^2 + c\zeta^2 = 0,$$

and so an arbitrary point $Q(\xi, \eta, \zeta, \tau + \lambda)$ of the line joining P to the vertex $T(0, 0, 0, 1)$ of the tetrahedron of reference also lies on the surface. The surface is therefore traced by a system of lines all passing through the point T. Such a surface is called a *cone*, and the point T is called the *vertex* of the cone. Note that T is the point of intersection of the three linearly independent planes in terms of which the equation of the cone is expressed, namely, $x = 0$, $y = 0$, $z = 0$.

In the plane $t = 0$, the equation

$$ax^2 + by^2 + cz^2 = 0$$

represents a conic Γ, and so the cone is generated by the lines joining the vertex T to the conic Γ. It will follow from subsequent work that *any* plane not through T cuts the cone in a non-degenerate conic.

DEFINITION. The cone, the plane-pair and the repeated plane are called *singular* quadrics. Otherwise the quadric is said to be *non-singular*.

ILLUSTRATION 1. *A plane through T cuts the cone in the points of two straight lines, which may 'coincide'.*

The equation of an arbitrary plane through T may be taken in the form

$$lx+my+nz = 0.$$

Suppose that (ξ, η, ζ, τ) is any point whatever, other than T itself, common to the plane and the cone. Then

$$a\xi^2+b\eta^2+c\zeta^2 = 0, \quad l\xi+m\eta+n\zeta = 0.$$

Eliminating ζ, we obtain a quadratic equation for ξ/η which has, in general, two solutions, say

$$\xi = \alpha\eta, \quad \xi = \beta\eta,$$

where α and β may be calculated in terms of a, b, c, l, m, n if desired. Thus the point (ξ, η, ζ, τ) lies in one or other of the planes

$$x = \alpha y, \quad x = \beta y$$

as well as in the given plane

$$lx+my+nz = 0.$$

It therefore lies on one or other of the lines in which these two planes meet the given plane.

The point $T(0, 0, 0, 1)$ lies in each of the three planes just named, and so each of the lines of intersection passes through T.

4. Notation. We write

$$S \equiv ax^2+by^2+cz^2+dt^2+2fyz+2gzx+2hxy+2uxt+2vyt+2wzt$$

and S_{11} for S with (x, y, z, t) replaced by (x_1, y_1, z_1, t_1).

We also write

$$X \equiv ax+hy+gz+ut = \frac{1}{2}\frac{\partial S}{\partial x},$$

$$Y \equiv hx+by+fz+vt = \frac{1}{2}\frac{\partial S}{\partial y},$$

$$Z \equiv gx+fy+cz+wt = \frac{1}{2}\frac{\partial S}{\partial z},$$

$$T \equiv ux+vy+wz+dt = \frac{1}{2}\frac{\partial S}{\partial t}$$

and X_1, Y_1, Z_1, T_1 for X, Y, Z, T with (x, y, z, t) replaced by (x_1, y_1, z_1, t_1).

Finally, we write

$$S_1 \equiv xX_1 + yY_1 + zZ_1 + tT_1$$
$$\equiv x_1X + y_1Y + z_1Z + t_1T,$$

and

$$S_{12} \equiv x_1X_2 + y_1Y_2 + z_1Z_2 + t_1T_2$$
$$\equiv x_2X_1 + y_2Y_1 + z_2Z_1 + t_2T_1$$
$$\equiv S_{21},$$

and observe that $S \equiv xX + yY + zZ + tT.$

We shall denote the *discriminant* of S by

$$\Delta \equiv \begin{vmatrix} a & h & g & u \\ h & b & f & v \\ g & f & c & w \\ u & v & w & d \end{vmatrix},$$

and we assume that Δ does not vanish unless the contrary is stated explicitly.

5. Joachimstal's equation. Let

$$P_1(x_1, y_1, z_1, t_1), \quad P_2(x_2, y_2, z_2, t_2)$$

be any two points. Then the condition that the point Q whose symbol is

$$\mathbf{Q} \equiv \lambda_1\mathbf{P}_1 + \lambda_2\mathbf{P}_2$$

lies on the quadric S reduces, on substituting and rearranging, to

$$\lambda_1^2 S_{11} + 2\lambda_1\lambda_2 S_{12} + \lambda_2^2 S_{22} = 0.$$

This is a *quadratic* equation for the ratio λ_1/λ_2 (or λ_2/λ_1) which determines the position of Q on the line, and therefore *a general straight line cuts the quadric in two points.*

Theorem-example. The curve in which an arbitrary plane cuts the quadric is met by a general line of the plane in two points.

Hence *the plane sections of a quadric are conics.*

6. Tangency. DEFINITION. A straight line which meets a quadric in one point only is said to be a *tangent* at that point or to *touch* the quadric there, and the point is called the *point of contact* of the tangent. It is convenient to regard the tangent as a line meeting the quadric in two 'coincident' points.

As the first properties of the tangents of quadrics are closely analogous to those of conics, we enunciate them in the form of theorem-examples to be solved by the reader. (See, for example, M., Chapter v.)

Theorem-examples. 1. The tangents to the quadric surface S at a given point P_1 all lie in the plane

$$S_1 = 0,$$

or
$$X_1x+Y_1y+Z_1z+T_1t=0.$$

This plane is called the *tangent plane* at P_1 to the quadric and P_1 is called the *point of contact.*

The tangent plane is, in general, uniquely defined. The case of exception, which we shall examine later, arises if it is possible for the coefficients X_1, Y_1, Z_1, T_1 to vanish simultaneously.

[See M., p. 74.]

2. The tangents to the quadric surface S which pass through a given point P_1 not on it generate the cone whose equation is

$$S_1^2 = S_{11}S.$$

This cone is called the *tangent cone* from P_1 to the quadric. If P_1 lies on S then the cone degenerates into the tangent plane at P_1 repeated.

[Compare M., p. 74.]

7. Conjugacy. DEFINITION. Two points P_1, P_2 are said to be *conjugate* with respect to a quadric S if the line P_1P_2 meets S in two points which separate P_1, P_2 harmonically.

Theorem-examples. 1. If P_1 and P_2 are conjugate with respect to S, then

$$S_{12} = 0.$$

[See M., p. 75.]

2. If P_1 is given, then the points P such that P_1 and P are conjugate with respect to S lie on the plane whose equation is

$$S_1 = 0.$$

This plane is called the *polar plane* of P_1 with respect to the quadric; P_1 itself is called the *pole* of its polar plane.

[Compare M., pp. 75–6.]

3. The polar plane of a point on a quadric is the tangent plane at that point.

[See M., p. 76.]

4. If the polar plane of P_1 passes through P_2, then the polar plane of P_2 passes through P_1.

[See M., p. 76.]

5. The polar plane of P_1 meets the quadric S in a conic every point of which is the point of contact of a tangent line from P_1 to the quadric.

[Compare M., p. 76.]

8. The condition for a plane to touch a quadric. Suppose that the plane

$$lx+my+nz+pt = 0$$

touches the quadric at the point (x_1, y_1, z_1, t_1). The equation of the tangent plane is

$$xX_1+yY_1+zZ_1+tT_1 = 0.$$

Since these two equations represent the same plane, their coefficients are proportional, and therefore there exists a number ρ, not zero, such that

$$ax_1+hy_1+gz_1+ut_1 = \rho l, \quad (1)$$

$$hx_1+by_1+fz_1+vt_1 = \rho m, \quad (2)$$

$$gx_1+fy_1+cz_1+wt_1 = \rho n, \quad (3)$$

$$ux_1+vy_1+wz_1+dt_1 = \rho p. \quad (4)$$

Further,

$$lx_1+my_1+nz_1+pt_1 = 0, \quad (5)$$

since (x_1, y_1, z_1, t_1) necessarily lies in its tangent plane. Eliminating $x_1:y_1:z_1:t_1:-\rho$, we obtain the condition that the plane should touch the quadric, in the form

$$\begin{vmatrix} a & h & g & u & l \\ h & b & f & v & m \\ g & f & c & w & n \\ u & v & w & d & p \\ l & m & n & p & 0 \end{vmatrix} = 0.$$

On expanding this determinant, we find that the condition becomes

$$Al^2+Bm^2+Cn^2+Dp^2+2Fmn+2Gnl+2Hlm$$
$$+2Ulp+2Vmp+2Wnp = 0,$$

where A, B, ... are the cofactors of a, b, ... in the determinant Δ.

ALITER. We can find this condition, and also the ratios

$$x_1 : y_1 : z_1 : t_1$$

which give the point of contact, as follows. Multiply the equations (1), (2), (3), (4) by A, H, G, U respectively and add. Then, using standard properties of determinants,

$$\Delta x_1 = \rho(Al + Hm + Gn + Up).$$

Similarly

$$\Delta y_1 = \rho(Hl + Bm + Fn + Vp),$$

$$\Delta z_1 = \rho(Gl + Fm + Cn + Wp),$$

$$\Delta t_1 = \rho(Ul + Vm + Wn + Dp).$$

On substituting these results in equation (5), we obtain the required condition.

9. Duality. The ideas of duality as they apply to quadrics are in many ways closely analogous to those for conics.

Consider the two equations

$$S \equiv ax^2 + by^2 + cz^2 + dt^2 + 2fyz + 2gzx + 2hxy + 2uxt + 2vyt + 2wzt = 0,$$

$$\Sigma \equiv a'l^2 + b'm^2 + c'n^2 + d'p^2 + 2f'mn + 2g'nl + 2h'lm + 2u'lp + 2v'mp + 2w'np = 0.$$

The first equation defines a system of points which we have just been studying as the points of a quadric S; the second equation defines a system of planes whose properties we now examine. The system of points is called a *locus* and the system of planes is called an *envelope*. The line joining two points P_1, P_2 of S is called the *chord* P_1P_2 of the locus, and the line of intersection of two planes π_1, π_2 of Σ is called the *axis* $(\pi_1\pi_2)$ of the envelope.

10. The equation of a quadric envelope as a 'sum of squares'. Just as in the case of point coordinates, the quadratic form Σ can be expressed as the sum of the squares of four linearly independent forms each of which is linear in l, m, n, p. The standard form is then

$$al^2 + bm^2 + cn^2 + dp^2 = 0,$$

or, by special choice of the unit plane,

$$l^2 + m^2 + n^2 + p^2 = 0.$$

We consider first the cases when one or more of the coefficients a, b, c, d vanish.

(i) *The case* $a = b = c = 0$. The equation is

$$p^2 = 0,$$

which represents a *degenerate tangential quadric* consisting of the 'repeated' point T, whose equation is $p = 0$.

(ii) *The case* $b = c = 0$. The equation is

$$al^2 + dp^2 = 0$$

or

$$(l\sqrt{a} + ip\sqrt{d})(l\sqrt{a} - ip\sqrt{d}) = 0,$$

which represents a degenerate tangential quadric consisting of the two distinct points $(\sqrt{a}, 0, 0, \pm i\sqrt{d})$, whose equations are

$$l\sqrt{a} \pm ip\sqrt{d} = 0.$$

They lie on the line joining the points X, T whose equations are $l = 0$, $p = 0$.

(iii) *The case* $d = 0$. The equation is

$$al^2 + bm^2 + cn^2 = 0,$$

which we must consider in more detail.

We note first that, if the plane $\pi(\lambda, \mu, \nu, \rho)$, whose equation is $\lambda x + \mu y + \nu z + \rho t = 0$, is a plane of the quadric envelope, then

$$a\lambda^2 + b\mu^2 + c\nu^2 = 0,$$

and so an arbitrary plane $\pi'(\lambda, \mu, \nu, \rho + k)$ through the line of intersection of π with the plane $t = 0$ also belongs to the envelope. The envelope consists therefore of a system of planes which can be defined as follows:

Take in the plane $t = 0$ *the conic envelope whose 'tangential equation' referred to the triangle* XYZ *is* $al^2 + bm^2 + cn^2 = 0$; *the planes of the quadric envelope are the planes which pass through the lines of the conic envelope.*

It is customary to say, loosely, that the quadric envelope has degenerated into the conic envelope; we shall use this language, while warning the reader to be quite clear about its full meaning.

The degeneration of a quadric envelope into a conic envelope is exactly analogous to the degeneration of a quadric locus into a cone. A cone is a system of points which lie on lines through a fixed point; the dual is a system of planes which pass through lines in a fixed plane.

DEFINITION. A quadric envelope which 'degenerates into a conic envelope' in this way is often called a *disc quadric*.

11. Tangency and conjugacy for a quadric envelope. We now take the equation of the envelope in the general form

$$\Sigma \equiv al^2+bm^2+cn^2+dp^2+2fmn+2gnl+2hlm$$
$$+2ulp+2vmp+2wnp = 0,$$

and use the notation Σ_1, Σ_{11}, L, M, N, P, L_1, M_1, N_1, P_1, etc., analogous to the notation S_1, S_{11}, X, Y, Z, T, X_1, Y_1, Z_1, T_1 for a quadric locus.

The two planes $\pi_1(l_1, m_1, n_1, p_1)$, $\pi_2(l_2, m_2, n_2, p_2)$ meet in a line; two planes of the envelope pass through this line, their coordinates $(\lambda_1 l_1+\lambda_2 l_2,\ \lambda_1 m_1+\lambda_2 m_2,\ \lambda_1 n_1+\lambda_2 n_2,\ \lambda_1 p_1+\lambda_2 p_2)$ being given by Joachimstal's equation

$$\lambda_1^2\Sigma_{11}+2\lambda_1\lambda_2\Sigma_{12}+\lambda_2^2\Sigma_{22} = 0.$$

Hence *two planes of the envelope pass through each axis.*

Suppose first that the plane π_1 is a plane of the envelope; then $\Sigma_{11} = 0$, so that one root of the equation in λ_2/λ_1 vanishes and one of the two planes coincides with π_1. If, in addition, we choose the plane π_2 so that $\Sigma_{12} = 0$, then each of the two planes through the axis $(\pi_1\pi_2)$ 'coincides' with π_1 itself. In other words, the axes $(\pi_1\pi_2)$ which lie in the plane π_1 of the envelope, and which are such that $\Sigma_{12} = 0$, have π_1 itself as the unique ('repeated') plane of Σ through them. But, from the condition $\Sigma_{12} = 0$, the coordinates of the planes π_2 satisfy the equation $\Sigma_1 = 0$, which is

$$lL_1+mM_1+nN_1+pP_1 = 0.$$

Hence the planes π_2 all pass through the point with coordinates (L_1, M_1, N_1, P_1). This point is called a *contact* of the envelope. From the way in which it was derived, or, alternatively, by the relation

$$l_1L_1+m_1M_1+n_1N_1+p_1P_1 \equiv \Sigma_{11} = 0,$$

the contact lies in the plane π_1. Hence *each plane π_1 of the envelope Σ contains a uniquely defined point, called a contact. If we have any axis through the contact and lying in π_1, then the two planes of the envelope through that axis coincide with π_1.* The axes through the contact may be called the *contact axes* in π_1.

It may help the reader over unfamiliar ground to quote these results in dual form:

Each point P_1 of a quadric locus S lies in a uniquely defined plane, called a tangent plane. If we have any chord in the tangent plane and passing through P_1, then the two points of the quadric on that chord coincide with P_1. The chords in the tangent plane are the *tangent lines* through P_1.

Suppose next that π_1 is a given plane which does not belong to the envelope Σ, so that $\Sigma_{11} \neq 0$. Through an arbitrary line of π_1 there pass two planes of the envelope. If we define that line by the intersection of π_1 with another plane π_2, then the two planes of the envelope passing through it are given by Joachimstal's equation

$$\lambda_1^2\Sigma_{11} + 2\lambda_1\lambda_2\Sigma_{12} + \lambda_2^2\Sigma_{22} = 0.$$

When the line is chosen to be a contact axis, these two planes coincide, and so

$$\Sigma_{11}\Sigma_{22} = \Sigma_{12}^2.$$

The planes π_2 defining the contact axes in π_1 have therefore coordinates (l_2, m_2, n_2, p_2) subject to the equation

$$\Sigma_{11}\Sigma - \Sigma_1^2 = 0. \qquad (1)$$

This is a quadratic equation, so the planes π_2 are the planes of a quadric envelope. But, from the very definition of the system, this envelope must 'degenerate into a conic envelope', precisely as in § 10 (iii). The equation (1) therefore defines a conic envelope in the plane π_1, touched by all the contact axes of Σ which lie in π_1. Hence *the contact axes which lie in a plane π_1 touch the degenerate quadric envelope given by the equation*

$$\Sigma_{11}\Sigma - \Sigma_1^2 = 0.$$

[Dually, the tangent lines through a point P_1 generate the cone given by the equation $S_{11}S - S_1^2 = 0$.]

DEFINITION. Two planes π_1, π_2 are said to be *conjugate* with respect to the quadric envelope Σ if they are separated harmonically by the two planes of the envelope which pass through the line $(\pi_1\pi_2)$.

Theorem-examples. 1. If π_1 and π_2 are conjugate with respect to Σ, then

$$\Sigma_{12} = 0.$$

2. If π_1 is given, then the planes π, such that π_1 and π are conjugate with respect to Σ, pass through the point whose equation is

$$\Sigma_1 = 0.$$

This point is called the *pole* of π_1 with respect to the quadric envelope; π_1 itself is called the *polar* of its pole.

3. The pole of a plane which belongs to the envelope is the contact in that plane.

4. If the pole of π_1 lies in π_2, then the pole of π_2 lies in π_1.

5. The condition that the point (x, y, z, t) should be a contact of the quadric envelope

$$al^2 + bm^2 + cn^2 + dp^2 + 2fmn + 2gnl + 2hlm + 2ulp + 2vmp + 2wnp = 0$$

is that it should lie on the quadric locus

$$Ax^2 + By^2 + Cz^2 + Dt^2 + 2Fyz + 2Gzx + 2Hxy + 2Uxt + 2Vyt + 2Wzt = 0,$$

where $A, B, \ldots$ are the cofactors of $a, b, \ldots$ in the determinant Δ. (Dualise § 8.)

12. Quadric locus and quadric envelope.

From the results of § 8 and of § 11, Theorem-example 5, the tangent planes of a quadric locus are the planes of a quadric envelope, and the contacts of a quadric envelope are the points of a quadric locus. A locus and an envelope so related are said to be *associated*.

Moreover, if S is a quadric locus and Σ its associated envelope, then the associated locus of Σ is S itself. For if S is given by the equation

$$S \equiv ax^2 + \ldots = 0,$$

then Σ is given by the equation

$$\Sigma \equiv Al^2 + \ldots = 0,$$

where $A, B, \ldots$ are the cofactors of $a, b, \ldots$ in the determinant

$$\Delta \equiv \begin{vmatrix} a & h & g & u \\ h & b & f & v \\ g & f & c & w \\ u & v & w & d \end{vmatrix} \neq 0.$$

The associated locus of Σ is likewise given by

$$S' \equiv a'x^2 + \ldots = 0,$$

where a', b', ... are the cofactors of A, B, ... in the determinant

$$\begin{vmatrix} A & H & G & U \\ H & B & F & V \\ G & F & C & W \\ U & V & W & D \end{vmatrix}.$$

Now it is a well-known theorem in the theory of determinants that

$$a' = \Delta^2 a, \quad b' = \Delta^2 b, \quad \ldots,$$

and so the equations $\quad S = 0, \quad S' = 0$

represent the same surface.

Similarly, if Σ is a quadric envelope and S its associated locus, then the associated envelope of S is Σ itself.

Theorem-example. Show that a point P has the same polar with respect to a quadric surface S and its associated envelope Σ; and that a plane π has the same pole with respect to Σ and S.

13. Reciprocation. The method of reciprocation and its leading properties are very similar to the corresponding results for conics, and we give merely a brief account of them.

Let R be a given non-singular quadric. Consider a configuration containing points such as P and planes such as λ. With respect to the quadric R, let

the polar plane of P be the plane π,

the pole of λ be the point L.

Then the configuration containing the planes π and the points L is called the *reciprocal* of the given configuration with respect to the quadric R. The reciprocal of the new configuration with respect to R is the given configuration itself.

14. The reciprocal of a straight line. Suppose that a range of points on a line is determined in terms of a parameter θ by means of the coordinates

$$(x_1 + \theta x_2, \; y_1 + \theta y_2, \; z_1 + \theta z_2, \; t_1 + \theta t_2)$$

of a typical point P. The polar of P with respect to the quadric

$$R = 0$$

is given by the equation $\quad R_1 + \theta R_2 = 0.$

The reciprocal of the range of points is therefore a pencil of planes, and the cross-ratio of four points of the range is equal to the cross-ratio of the four corresponding planes of the pencil, each being the cross-ratio of the four values of θ which determine them.

It is customary to say, somewhat loosely, that *the reciprocal of a line with respect to a quadric is a line.* By this, we really mean that the reciprocal of the range of points on either line is the pencil of planes through the other.

Theorem-examples. 1. The reciprocal of the point (lm) at which two lines l, m meet is a plane $[l'm']$ in which the reciprocal lines lie.

2. Two skew lines reciprocate into two skew lines.
[Consider what would happen if the reciprocals had a common point.]

15. Analytical treatment of reciprocation. Let the equation of R be

$$R \equiv ax^2 + by^2 + cz^2 + dt^2 + 2fyz + 2gzx + 2hxy + 2uxt + 2vyt + 2wzt = 0.$$

Then the reciprocal of a point $P(\xi, \eta, \zeta, \tau)$ is the plane π whose plane-coordinates $(\lambda, \mu, \nu, \rho)$ are given by the relations

$$\begin{aligned} \lambda &= a\xi + h\eta + g\zeta + u\tau, \\ \mu &= h\xi + b\eta + f\zeta + v\tau, \\ \nu &= g\xi + f\eta + c\zeta + w\tau, \\ \rho &= u\xi + v\eta + w\zeta + d\tau. \end{aligned}$$

If we denote the determinant of the coefficients in R, which is not zero since R is non-singular, by Δ and the cofactors of $a, b, \ldots$ in Δ by $A, B, \ldots$, then the four equations can be solved to give ξ, η, ζ, τ in terms of λ, μ, ν, ρ in the form

$$\begin{aligned} \Delta\xi &= A\lambda + H\mu + G\nu + U\rho, \\ \Delta\eta &= H\lambda + B\mu + F\nu + V\rho, \\ \Delta\zeta &= G\lambda + F\mu + C\nu + W\rho, \\ \Delta\tau &= U\lambda + V\mu + W\nu + D\rho. \end{aligned}$$

Deduce from these equations the following results:

Theorem-examples. 1. The reciprocals of the points of a line are the planes through a line, and conversely. (See also § 14.)

2. The reciprocals of the points of a quadric locus are the planes of a quadric envelope, and conversely.

We say, loosely, that *the reciprocal of a quadric S with respect to a quadric R is a quadric S'.*

16. The properties of a cone. The properties of a cone in space are closely analogous to those of a line-pair in a plane, and we confine ourselves to stating the results in the form of a number of Theorem-examples. (Compare M., pp. 81–4.)

Theorem-examples. 1. If the equation

$$S \equiv ax^2+by^2+cz^2+dt^2+2fyz+2gzx+2hxy+2uxt+2vyt+2wzt = 0$$

represents a cone whose vertex is $P_1(x_1, y_1, z_1, t_1)$, then, in the notation of § 4,

$$X_1 = 0, \quad Y_1 = 0, \quad Z_1 = 0, \quad T_1 = 0.$$

[If P_2 is an arbitrary point of space, Joachimstal's equation derived from P_1P_2 shows that $S_{12} \equiv 0$ for all positions of P_2.]

2. The condition that the equation $S = 0$ should represent a cone is

$$\Delta \equiv \begin{vmatrix} a & h & g & u \\ h & b & f & v \\ g & f & c & w \\ u & v & w & d \end{vmatrix} = 0.$$

[An immediate consequence of the preceding.]

3. Conversely, if $\Delta = 0$, then the equation $S = 0$ represents a cone. [In special cases the cone may itself degenerate into a plane pair or into a repeated plane.]

4. The polar of an arbitrary point passes through P_1.

5. The polar of P_1 itself is indeterminate.

6. The pole of a plane not through P_1 is the point P_1.

7. A plane through P_1 has an infinite number of poles, all of which lie on a straight line through P_1.

17. Self-polar tetrahedron. Let S be a given quadric. We propose to establish the existence of tetrahedra in which each vertex is the pole of the opposite face, and to show that, referred to such a *self-polar tetrahedron* (as it is called) the equation of S can be taken in the form $ax^2+by^2+cz^2+dt^2 = 0$. We shall do this by successive choice of the vertices X, Y, Z, T.

In the first instance, the equation of S assumes the general form

$$S \equiv ax^2+by^2+cz^2+dt^2+2fyz+2gzx+2hxy$$
$$+2uxt+2vyt+2wzt = 0.$$

Suppose now that the tetrahedron has been chosen so that Y, Z, T all lie in the polar plane of the point (not on S) taken to be X. The pairs of points (X, Y), (X, Z), (X, T) are therefore each conjugate with respect to the quadric, and so $h = 0$, $g = 0$, $u = 0$.

Suppose next that Y has been chosen arbitrarily in the plane (not on S) and that Z, T lie in the polar plane of Y. Then (Y, Z), (Y, T) are also conjugate pairs, and so $f = 0$, $v = 0$.

The quadric now is $ax^2+by^2+cz^2+dt^2+2wzt = 0$, and the line common to the planes $x = 0$, $y = 0$ meets it where $cz^2+dt^2+2wzt = 0$. This is not satisfied identically, otherwise we should have $c = d = w = 0$, and the quadric would consist merely of two planes. Hence, solving for the ratio $z:t$, we find that the line meets the quadric in two points U, V. Choose Z arbitrarily on the line, not at U or V (that is, not on S), and take T in the polar plane of Z. Then (Z, T) is a conjugate pair, and so $w = 0$.

We have therefore defined a self-polar tetrahedron as required, and shown incidentally that the equation of S referred to it is of the form

$$ax^2+by^2+cz^2+dt^2 = 0.$$

Theorem-examples. 1. Each of the triangles YZT, ZXT, XYT, XYZ is self-polar with respect to the conic in which its plane cuts S.

2. The tangential equation of the quadric $ax^2+by^2+cz^2+dt^2 = 0$ is

$$l^2/a+m^2/b+n^2/c+p^2/d = 0.$$

[We shall frequently use this expression to mean 'the equation in tangential coordinates of the associated quadric envelope'.]

ILLUSTRATION 2. *The poles of the faces YZT, ZXT, XYT, XYZ of the tetrahedron of reference with respect to the quadric envelope*

$$\Sigma \equiv al^2+bm^2+cn^2+dp^2+2fmn+2gnl+2hlm$$
$$+2ulp+2vmp+2wnp = 0$$

are P, Q, R, S respectively. Prove that, if XP, YQ, ZR, TS are concurrent, then

$$fu = gv = hw.$$

The equation of the pole of the plane $(1, 0, 0, 0)$ is

$$al + hm + gn + up = 0,$$

and so $P \equiv (a, h, g, u)$. The coordinates of any point of XP are therefore

$$(\lambda, h, g, u).$$

Points similarly defined on YQ, ZR, TS are

$$(h, \mu, f, v),$$
$$(g, f, \nu, w),$$
$$(u, v, w, \rho).$$

If the lines are concurrent, these four points can be chosen to be the same, and so (taking ratios from those coordinates which do not involve the unknowns λ, μ, ν, ρ) we have

$$fu = gv = hw.$$

ILLUSTRATION 3. *S is a given quadric, π an arbitrary plane and P_1 an arbitrary point. To prove that the cone projecting the conic $C \equiv (S\pi)$ from P_1 meets S again in a conic, and that the equation of the plane of the conic is $S_{11}\pi - 2\pi_1 S_1 = 0$.*

Let P_2 be a point of C, and let $P_1 P_2$ meet S again in P_3. Then there exists a relation

$$\mathbf{P}_2 \equiv \lambda \mathbf{P}_1 + \mu \mathbf{P}_3.$$

Since P_2 lies on C, we have (in obvious notation)

$$S_{22} = 0, \quad \pi_2 = 0,$$

and also, since P_3 lies on S,

$$S_{33} = 0.$$

Hence the relation

$$S_{22} \equiv \lambda^2 S_{11} + 2\lambda\mu S_{13} + \mu^2 S_{33}$$

gives

$$\lambda S_{11} + 2\mu S_{13} = 0.$$

Further,

$$\pi_2 \equiv \lambda\pi_1 + \mu\pi_3,$$

so that

$$\lambda\pi_1 + \mu\pi_3 = 0.$$

Eliminating $\lambda : \mu$, we have

$$S_{11}\pi_3 - 2\pi_1 S_{13} = 0,$$

and so P_3, as it varies, lies in the plane whose equation is

$$S_{11}\pi - 2\pi_1 S_1 = 0.$$

Note. This plane passes through the line in which π meets the polar plane of P_1.

EXAMPLES II

1. Prove that the equation

$$y^2+z^2+2yz+2zx+2xy+2xt+2yt+2zt = 0$$

represents a cone whose vertex is the point $(0, 1, -1, 0)$, and verify that the polar of an arbitrary point (ξ, η, ζ, τ) passes through the vertex.

2. Prove that the equation

$$l^2-m^2-n^2+p^2+2mn+2lp = 0$$

is the tangential equation of a pair of points.

3. Prove that, as λ varies, the equation

$$(ax^2+by^2+cz^2+dt^2)+\lambda(x^2+y^2+z^2+t^2) = 0,$$

where a, b, c, d are unequal, represents a system of quadrics of which four are cones. Discuss the cases (i) $a = d$, $b \neq c$; (ii) $a = d$, $b = c$.

4. Find the equation of the tangent cone from the point $(1, 1, 1, 1)$ to the quadric

$$ax^2+by^2+cz^2+dt^2 = 0.$$

5. Prove that the polar planes of the points of the quadric

$$ax^2+by^2+cz^2+dt^2 = 0$$

with respect to the quadric

$$x^2+y^2+z^2+t^2 = 0$$

are the planes of the quadric envelope

$$al^2+bm^2+cn^2+dp^2 = 0.$$

6. Prove that the four harmonic inverses of a point of a quadric, with respect to the vertices and opposite faces of a self-polar tetrahedron, all lie on the quadric.

7. The transversals from an arbitrary point U meet the edges YZ, ZX, XY, XT, YT, ZT of a tetrahedron in L, M, N, P, Q, R respectively. Prove that a quadric can be drawn to touch the edges of the tetrahedron at these points. Identify the polar planes of X, Y, Z, T, and prove that the tangent plane at L passes through the harmonic conjugate of P with respect to X and T.

8. In Ex. 7, the poles of the planes YZT, ZXT, XYT, XYZ with respect to the quadric are X', Y', Z', T'. Prove that the lines XX', YY', ZZ', TT' all pass through U.

9. Prove that the quadric $2fyz+2uxt = 0$ contains the four lines XY, XZ, TY, TZ. Prove also that, for all values of the ratio f/u, the polar plane, with respect to this quadric, of the point (ξ, η, ζ, τ) passes through the line joining the points $(\xi, 0, 0, -\tau)$, $(0, \eta, -\zeta, 0)$.

10. The lines joining an arbitrary point P to the vertices X, Y, Z, T of a tetrahedron meet the opposite faces in A, B, C, D respectively. Prove that there exists a quadric touching the faces of the tetrahedron at these points.

11. The equations of a given quadric and a given plane are

$$ax^2+by^2+cz^2+dt^2 = 0, \quad lx+my+nz+pt = 0$$

respectively. Prove that the equation

$$p^2(ax^2+by^2+cz^2)+d(lx+my+nz)^2 = 0$$

represents a cone, that the vertex of this cone is the point $T(0, 0, 0, 1)$, and that the cone passes through the conic in which the plane and quadric intersect.

12. Prove that the equation of the cone whose vertex is the point $X(1, 0, 0, 0)$ and which passes through the conic in which the plane

$$lx+my+nz+pt = 0$$

meets the quadric $xt-yz = 0$ is

$$(my+nz+pt)\,t+lyz = 0.$$

13. Prove that the tetrahedron whose vertices are the points $(1, 0, 0, 0)$, $(0, 1, 2, 3)$, $(0, 1, 1, -1)$, $(0, -4, 5, 1)$ is self-polar with respect to the quadric $x^2+t^2+2yz = 0$.

14. Prove that, whatever the value of λ, the line whose equations are $x = \lambda y$, $z = \lambda t$ lies on the quadric $xt = yz$.

15. Prove that the tangent plane to the quadric $xt = yz$ at the point $(8, 4, 2, 1)$ meets the quadric in a degenerate conic, consisting of the two lines in which the tangent plane is met by the planes $x = 2y$ and $x = 4z$.

16. Find the coordinates of the points in which the line joining the points $(1, 1, 1, 1)$ and $(3, 1, -1, -3)$ meets the tetrahedron of reference. Find the polar planes of these four points with respect to the quadric

$$x^2+y^2+z^2+t^2+2xt = 0;$$

prove that the four planes so obtained belong to a pencil, and verify by direct calculation that the cross-ratios of the four points and the four corresponding planes are equal.

17. Prove that the four planes

$$x = 0, \quad x+y = 0, \quad x+y+z = 0, \quad x+y+z+t = 0$$

form a tetrahedron self-polar with respect to the quadric

$$4x^2+3y^2+2z^2+t^2+4yz+4zx+6xy+2xt+2yt+2zt = 0.$$

18. Prove that the tangential equation of the quadric $x^2+y^2+z^2+2xt = 0$ is $m^2+n^2-p^2+2lp = 0$, and that the pole of the plane $x+y+z+t = 0$ is the point $(1, 1, 1, 0)$.

MISCELLANEOUS EXAMPLES II

1. Prove that the equation of a quadric which touches XZT and YZT at X and Y respectively is

$$cz^2 + dt^2 + 2hxy + 2wzt = 0.$$

Prove that, if through a point P on this quadric a line can e drawn touching the quadric and meeting the edges XZ and YT, then I must lie in one of two fixed planes through XY. [P.]

2. Show that every quadric of the system

$$apx^2 + ary^2 + bpz^2 + brt^2 + 2hqyz + 2hpzx + 2aqxy + 2hqxt + 2hryt + 2bqzt = 0$$

that satisfies the condition for a cone, the a, h, b, p, q, r being otherwise arbitrary, is a plane pair. [M.T. II.]

3. Find the equation of the quadric S' which is the reciprocal of the quadric

$$S \equiv x^2 + y^2 + z^2 + t^2 = 0$$

with respect to the non-singular quadric

$$R \equiv ax^2 + by^2 + cz^2 + dt^2 = 0.$$

The polar of the unit point $(1, 1, 1, 1)$ with respect to R is the plane π. The cone whose vertex is $X(1, 0, 0, 0)$ and which passes through the conic (S, π) meets S again in a conic whose plane is π_1, and planes π_2, π_3, π_4 are defined similarly from the vertices Y, Z, T of the tetrahedron of reference. Prove that the tetrahedron whose faces are π_1, π_2, π_3, π_4 is self-polar with respect to S'. [M.T. II.]

4. Points P, Q, R, L, M, N are taken on the edges YZ, ZX, XY, XT, YT, ZT respectively of a tetrahedron $XYZT$, so that the planes YZL, ZXM, XYN, XTP, YTQ, ZTR are concurrent. Prove that the quadrics which pass through Q, R, M, N and touch ZX, XY and YT also touch ZT.

Prove also that these quadrics include three systems of cones:

(i) a system with vertex at the harmonic conjugate of L with respect to X, T;

(ii) a system with vertex at the harmonic conjugate of P with respect to Y, Z;

(iii) a system whose vertices all lie on the line LP. [M.T. II.]

CHAPTER III

THE GENERATORS OF A QUADRIC SURFACE

1. A simple form for the equation of a quadric. In Chapter II, § 3, we obtained the equation of a non-singular quadric in the form

$$ax^2 + by^2 + cz^2 + dt^2 = 0.$$

Write this equation in the alternative form

$$(x\sqrt{a} + it\sqrt{d})(x\sqrt{a} - it\sqrt{d}) + (y\sqrt{b} + iz\sqrt{c})(y\sqrt{b} - iz\sqrt{c}) = 0,$$

and make the transformation of coordinates

$$x' = x\sqrt{a} + it\sqrt{d}, \quad t' = x\sqrt{a} - it\sqrt{d},$$
$$y' = y\sqrt{b} + iz\sqrt{c}, \quad z' = -y\sqrt{b} + iz\sqrt{c},$$

so that the equation becomes

$$x't' = y'z'.$$

Dropping dashes, we obtain the equation of a non-singular quadric in the simple form

$$xt = yz.$$

We proceed to investigate the properties of a quadric S given by this equation.

2. The two systems of generators. Consider the system of straight lines given, for varying values of λ, by the equations

$$x = \lambda z, \quad y = \lambda t.$$

The coordinates of any point of this line satisfy the equation $xt = yz$, whatever the value of λ may be. Hence *there is on the quadric surface S an infinite system of straight lines, the individual lines of the system being defined by the values of the parameter λ.*

There is likewise a *second system of straight lines* given, for varying values of μ, by the equations

$$x = \mu y, \quad z = \mu t.$$

Any straight line lying on a quadric surface is called a *generator*. We have succeeded in obtaining two systems of generators, to be called *the λ-system* and *the μ-system* respectively. Each of these systems of generators is also called a *regulus*, and the two reguli are said to be *complementary*.

Theorem-examples. 1. Two distinct generators of the λ-system do not intersect, and two distinct generators of the μ-system do not intersect.

[Try solving the equations $x = \lambda_1 z$, $y = \lambda_1 t$; $x = \lambda_2 z$, $y = \lambda_2 t$.]

2. Each generator of the λ-system meets each generator of the μ-system.

3. One generator of each system passes through each point of the quadric.

4. The point of intersection of the two generators with parameter-values λ, μ respectively is $(\lambda\mu, \lambda, \mu, 1)$. This gives a *parametric representation* for the coordinates of the points of the quadric.

5. If a $(1, 1)$ correspondence is set up between the generators of the λ-system and the generators of the μ-system, expressed by means of the equation

$$a\lambda\mu + b\lambda + c\mu + d = 0,$$

then corresponding generators meet on the intersection of the quadric by the plane

$$ax + by + cz + dt = 0.$$

[Use the preceding result.]

6. If a straight line meets the quadric surface in three points, it lies entirely upon it.

[Use Joachimstal's equation.]

7. Every straight line lying upon the quadric belongs to one or other of the two systems of generators defined in the text; in other words, there are precisely *two* systems of generators on the quadric.

[If the point whose coordinates are

$$(\lambda_1\mu_1 + k\lambda_2\mu_2, \quad \lambda_1 + k\lambda_2, \quad \mu_1 + k\mu_2, \quad 1 + k)$$

lies on the quadric $xt = yz$, then

$$(\lambda_1 - \lambda_2)(\mu_1 - \mu_2) = 0.]$$

3. The tangent plane at a point. Let $P(\xi, \eta, \zeta, \tau)$ be a point of the quadric

$$S \equiv xt - yz = 0,$$

so that

$$\xi\tau - \eta\zeta = 0.$$

The tangent plane at P is given by the equation

$$\tau x - \zeta y - \eta z + \xi t = 0.$$

In particular, if P is the point $(\lambda_0\mu_0, \lambda_0, \mu_0, 1)$, then the equation of the plane is

$$x - \mu_0 y - \lambda_0 z + \lambda_0\mu_0 t = 0.$$

This plane meets the quadric in the points whose coordinates are of the form $(\lambda\mu, \lambda, \mu, 1)$, where, on substitution,

$$\lambda\mu - \mu_0\lambda - \lambda_0\mu + \lambda_0\mu_0 = 0.$$

Hence $$(\lambda - \lambda_0)(\mu - \mu_0) = 0.$$

The tangent plane at P therefore meets the quadric in points which are *either* the points $(\lambda_0\mu, \lambda_0, \mu, 1)$, for varying μ, of the λ-generator $\lambda = \lambda_0$, *or* the points $(\lambda\mu_0, \lambda, \mu_0, 1)$, for varying λ, of the μ-generator $\mu = \mu_0$. Hence *the tangent plane at a point P of the quadric meets the surface in the two generators through P.*

4. The projective generation of a quadric surface. There are certain fundamental constructions which serve to define quadric surfaces, and the reader should be thoroughly familiar with all of them.

(i) *The lines of intersection of corresponding planes of two related pencils, whose axes are skew lines, form a regulus.* Take a tetrahedron of reference in which the two skew lines are respectively

$$x = 0, \quad t = 0$$

and $$y = 0, \quad z = 0.$$

A plane of the first pencil is

$$x + \lambda t = 0,$$

and a plane of the second pencil is

$$y + \mu z = 0,$$

where, in virtue of the $(1, 1)$ correspondence between the pencils, there is a relation of the form

$$a\lambda\mu + b\lambda + c\mu + d = 0.$$

Hence, for a point (x, y, z, t) common to two corresponding planes,

$$axy - bzx - cyt + dzt = 0,$$

and this is the equation of a quadric on which the variable lines are the generators of one system.

COROLLARY. The two given skew lines are members of the complementary regulus.

(ii) *The lines joining corresponding points of related ranges on two skew lines form a regulus.* Take a tetrahedron of reference in which the two skew lines are respectively

$$x = 0, \quad t = 0$$

and

$$y = 0, \quad z = 0.$$

The coordinates of corresponding points can be taken in the form $P(0, \lambda, 1, 0)$, $Q(\mu, 0, 0, 1)$, where, in virtue of the $(1, 1)$ correspondence between the ranges, there is a relation of the form

$$a\lambda\mu + b\lambda + c\mu + d = 0.$$

Now the coordinates of an arbitrary point of the line PQ are given by

$$\rho x = \mu, \quad \rho y = p\lambda, \quad \rho z = p, \quad \rho t = 1,$$

so that

$$\mu = x/t, \quad \lambda = y/z.$$

The line PQ therefore generates the quadric surface whose equation is

$$axy + byt + czx + dzt = 0.$$

Corollary. The two given skew lines are members of the complementary regulus.

(iii) *The lines which meet each of three given skew lines form a regulus.* Let a, b, c be the given skew lines; we propose to prove that the variable lines cut b and c in points which are related in a $(1, 1)$ correspondence, which is clearly algebraic. Let P be an arbitrary point of b. From P can be drawn a unique transversal to a and c (it is therefore a line of the given system), and this line cuts c in a unique point Q. Conversely, if Q is given on c, the unique transversal can be drawn from it to a and b. Hence P gives rise to a unique point Q, while Q arises from a unique point P; and so, by (ii), the line PQ generates a regulus.

Corollary. The three given skew lines are members of the complementary regulus.

Theorem-examples. 1. If two conics in different planes have two distinct points X, T common, then a tetrahedron of reference can be found (with a suitable unit point) so that the equations of the conics are respectively

$$z = 0, \quad y^2 = xt$$

and

$$y = 0, \quad z^2 = xt.$$

[Use the form for the equation of a conic given in M., p. 99.]

2. If there is a (1, 1) correspondence, between the points $(\theta^2, \theta, 0, 1)$ and $(\phi^2, 0, \phi, 1)$ of Theorem-example 1, in which X and T are each self-corresponding, then the equation of the correspondence can be taken in the form $\phi = k\theta$.

[See M., p. 32.]

3. The lines joining corresponding points in Theorem-example 2 generate the quadric surface

$$(ky+z)(y+kz) = kxt.$$

4. Four skew lines have two common transversals (which may, in special cases, 'coincide').

[Consider the quadric surface defined by three of them.]

5. If four skew lines have only one transversal, then each touches the quadric surface determined by the other three.

5. Polar lines. Let $S = 0$ be the equation of a given quadric and P_1, P_2 two given points whose join is a given line l. The symbol of an arbitrary point of l can be taken in the form $\mathbf{P} \equiv \lambda\mathbf{P}_1 + \mu\mathbf{P}_2$, from which it follows that the equation of the polar plane of P with respect to S can be put in the form

$$\lambda S_1 + \mu S_2 = 0.$$

As P varies on l, the ratio λ/μ varies, but the polar plane of P always passes through the line l' whose equations are

$$S_1 = 0, \quad S_2 = 0.$$

The line l' is called the *polar line* of l with respect to S.

Theorem-examples. 1. If P is any point of l and P' any point of l', then the points P and P' are conjugate with respect to S.

2. The polar line of l' is l.

3. If l meets the quadric in A, B and l' meets the quadric in A', B', then AA', BB' are generators of one system and AB', BA' are generators of the other system.

[The tangent plane at A meets the quadric in two generators; similarly for the other points.]

4. Conversely, if A and B are two points of the quadric and the λ-generator at A meets the μ-generator at B in A', while the μ-generator at A meets the λ-generator at B in B', then AB and $A'B'$ are polar lines.

5. If a line l is a tangent to S, then its polar line meets it at the point of contact. Conversely, if a line meets its polar line, then the plane which contains them is the tangent plane at their common point.

6. A generator of S is its own polar line. Conversely, if a line is its own polar line, then it is a generator of S. In other words, *a generator of S is self-reciprocal with respect to S.*

7. The opposite edges of a tetrahedron self-conjugate with respect to a quadric S are polar lines with respect to S.

6. Conjugate lines.

Let l be a given line and l' its polar line with respect to S. If m is any line meeting l', then l and m are said to be *conjugate* lines. We prove that *the polar line m' of m then meets l.* In fact, if m meets l' in P, then the polar plane of P, considered as a point of l', contains l, and the polar plane of P, considered as a point of m, contains m'. The lines l and m' therefore lie in the polar plane of P, and so they intersect.

ILLUSTRATION 1. *To find the conditions that the lines $x = 0, t = 0$ and $y = 0, z = 0$ should be polar lines with respect to the general quadric*
$S \equiv ax^2 + by^2 + cz^2 + dt^2 + 2fyz + 2gzx + 2hxy + 2uxt + 2vyt + 2wzt = 0.$

Since YZ and XT are polar lines, the pairs of points (Y, X), (Y, T), (Z, X), (Z, T) are all conjugate with respect to S; the conditions for this are

$$h = 0, \quad v = 0, \quad g = 0, \quad w = 0.$$

COROLLARY. If two pairs of opposite edges of a tetrahedron are polar lines with respect to a quadric, so also are the third pair, and the tetrahedron is self-polar with respect to the quadric.

ILLUSTRATION 2. *To find the condition that the lines $x = 0, t = 0$ and $y = 0, z = 0$ should be conjugate lines with respect to the quadric S.*

The polar line of the line $x = 0, t = 0$ is the line of intersection of the polar planes of the points $(0, 1, 0, 0)$, $(0, 0, 1, 0)$ on it, and the equations of these two planes are

$$hx + by + fz + vt = 0,$$
$$gx + fy + cz + wt = 0.$$

These two planes meet the line $y = 0, z = 0$ in the points

$$(v, 0, 0, -h), \quad (w, 0, 0, -g)$$

respectively. But the condition that the two given lines should be conjugate is that these two points should be the same, and this is

$$gv = hw.$$

COROLLARY. If two pairs of opposite edges of a tetrahedron are conjugate with respect to a quadric, so also are the third pair.

ILLUSTRATION 3. *Alternative proof of Von Staudt's theorem* (Chapter I, § 16, p. 15).

LEMMA. *There is a unique quadric which has a given tetrahedron as a self-polar tetrahedron and which has a given line in arbitrary position as a generator.* Take the tetrahedron as tetrahedron of reference, so that the equation of the quadric (if any) is of the form

$$ax^2+by^2+cz^2+dt^2 = 0.$$

A unique quadric of the system can be drawn through three given points (x_1, y_1, z_1, t_1), (x_2, y_2, z_2, t_2), (x_3, y_3, z_3, t_3), its equation being

$$\begin{vmatrix} x^2 & y^2 & z^2 & t^2 \\ x_1^2 & y_1^2 & z_1^2 & t_1^2 \\ x_2^2 & y_2^2 & z_2^2 & t_2^2 \\ x_3^2 & y_3^2 & z_3^2 & t_3^2 \end{vmatrix} = 0.$$

These three points may be chosen on the given line (in general position) without changing the argument—in particular, it can be verified that the equation does not vanish identically—and the quadric, which meets that line in three points, contains it entirely.

Von Staudt's theorem follows almost immediately. Suppose that a given line l meets the faces of a tetrahedron $XYZT$ in points (symmetrically named) A, B, C, D. Reciprocate with respect to the quadric which has $XYZT$ as a self-polar tetrahedron and l as a generator. Then l reciprocates into itself. Also the point A in which l meets the plane YZT reciprocates into the plane α which contains l and the point X. Hence the range A, B, C, D reciprocates into the pencil of planes $[lX]$, $[lY]$, $[lZ]$, $[lT]$, and the cross-ratio of the range is equal to the cross-ratio of the pencil by the theorem of Chapter II, § 14, p. 45.

ILLUSTRATION 4. *The reciprocation of one quadric into another.*

We proved in Chapter II that the reciprocal of a quadric S with respect to a quadric R is a quadric S'. We now consider the effect of reciprocation on the generators of S.

The lines of the λ-system reciprocate into lines which we shall call lines of the λ'-system, and similarly the lines of the μ-system reciprocate into lines of the μ'-system. Now the point $(\lambda\mu)$ reciprocates into the plane $[\lambda'\mu']$, so that each line of the λ'-system meets each line of the μ'-system and the lines λ', μ' thus form the two complementary reguli on the quadric S'.

There is clearly a $(1, 1)$ algebraic correspondence between the reciprocal generators λ, λ' and between the reciprocal generators μ, μ'.

EXAMPLES III

1. Prove (do not merely verify) that the equation of the quadric on which the three lines $x = 0$, $t = 0$; $y = 0$, $z = 0$; $x+y-z = 0$, $y+z-t = 0$ are generators is

$$zx+xy+yt-zt = 0.$$

Obtain the equations of the two generators through the point $(1, 1, 3, 2)$ in the form

$$2x = t,\ 3y = z \quad \text{and} \quad 2x+y-z = 0,\ y+z-2t = 0.$$

Find also the equation of the tangent plane at that point, and verify that it contains the two generators.

2. Prove that the equation of the quadric which has the lines $x = 0$, $y = 0$ and $x = 0$, $z = 0$ as generators, contains the conic $x^2-yz = 0$, $t = 0$ and passes through the point $(1, 0, 0, 1)$ is

$$x^2-yz-xt = 0.$$

Prove also that the equations of the two generators through the point $(\theta, \theta^2, 1, 0)$ of the conic are

$$y = \theta x, \quad x-t = \theta z$$

and

$$x = \theta z, \qquad y = \theta(x-t).$$

3. Prove that the two generators of the quadric

$$yz+zx+xy+xt+yt+zt = 0$$

which pass through the point $(1, 0, 0, 0)$ are given by the equations

$$\frac{y}{\omega^2} = \frac{z}{\omega} = \frac{t}{1}$$

and

$$\frac{y}{\omega} = \frac{z}{\omega^2} = \frac{t}{1},$$

where ω is a complex cube root of unity.

4. Prove that the two lines which meet each of the four lines

$$x = 0, t = 0; \quad y = 0, z = 0; \quad x+y = 0, z+t = 0;$$
$$x+2y+3z = 0, \ 3y+2z+t = 0$$

are given by the equations

$$x = t, y = z \quad \text{and} \quad x = -t, y = -z.$$

[Obtain your answers; do not merely verify.]

5. Prove that, if l, m, n are three given skew lines, then the harmonic inverse of each of them with respect to the other two lies on the quadric defined by the three lines.

MISCELLANEOUS EXAMPLES III

1. The four points A, B, C, D are joined respectively to A', B', C', D', the poles of the planes BCD, CAD, ABD, ABC with respect to a quadric Σ. Prove that the lines AA', BB', CC', DD' are in general generators of one system of a quadric. Examine the particular cases in which (i) AB, CD are *conjugate* with respect to Σ, i.e. the polar line of each meets the other, and (ii) AB, CD are conjugate and so also are CA, BD. [P.]

2. If the six sides of a skew hexagon are all generators of the same quadric, prove that the three lines which join the pairs of opposite vertices are concurrent and that the three lines which are intersections of tangent planes at pairs of opposite vertices are coplanar.

$ABCDEFGH$ is a skew octagon whose sides are all generators of the same quadric. Prove that the eight lines AD, DG, GB, BE, EH, HC, CF, FA are all generators of another quadric. [P.]

3. A fixed quadric has the four sides of a skew quadrilateral as generators; a variable line λ meeting the diagonals of this quadrilateral cuts the quadric in A, B and the polar line of λ with respect to the quadric cuts the quadric in C, D. Prove that, for all positions of λ, the tetrahedron $ABCD$ is self-conjugate with respect to another fixed quadric having the four sides of the skew quadrilateral as generators. [P.]

4. Prove that, if the polar line of a line λ with respect to a quadric meets another line μ, then the polar line of μ with respect to the quadric meets λ.

Find the envelope of a plane which cuts two fixed planes in two lines each of which meets the polar line of the other with respect to a quadric. [P.]

5. Find the equation of the surface generated by lines l which meet the three given lines

$$x = 0, \quad y+z+at = 0,$$
$$y = 0, \quad z+x+bt = 0,$$
$$z = 0, \quad x+y+ct = 0.$$

Show that the equations of the lines l are

$$ax+by+cz+\lambda t = 0,$$

$$a(b+c)x+b(c+a)y+a(b+c)z+abct = \lambda(x+y+z)$$

for different values of λ. [M.T. I, modified.]

6. A line through a point P meets a quadric S in the points A_1, A_2. The generators of the two systems on S through A_1, A_2 are g_1, g_2 and g_1', g_2'. Show that the points g_1g_2' and g_2g_1' lie in the polar plane of P with respect to S.

If g_1, g_2, g_3, g_4 are four generators of one system on S, if g_1', g_2', g_3', g_4' are four generators of the other system, and if the four points $g_1g_1', g_2g_2', g_3g_3', g_4g_4'$ lie in a plane π, prove that the remaining twelve intersections of the lines g_1, g_2, g_3, g_4 with the lines g_1', g_2', g_3', g_4' lie in sets of four on three planes. Show further that these three planes and π form a tetrahedron which is self-polar with respect to S. [P.]

7. A $(1, 1)$ correspondence is established between the generators of one system on a quadric and the generators of the opposite system. Find the locus of points of intersection of corresponding generators. [P.]

8. O is a point in general position with regard to a tetrahedron $ABCD$. The transversal from O to BC, AD meets BC in L and AD in P. Points M, Q and N, R are defined similarly on CA, BD and AB, CD. Prove that there is a quadric having MR, NP, LQ and NQ, LR, MP as sets of generators of opposite systems, and that this quadric cuts the plane ABC in a conic which touches BC at L, CA at M and AB at N. [P.]

9. The coordinates of any point on the quadric $xt = yz$ are given by $(\lambda\mu, \lambda, \mu, 1)$. Find the cross-ratio of the four points in which the line $z = 0$, $t = 0$ is met by the faces of the tetrahedron formed by the four points whose parameters are (λ_1, μ_1), (λ_2, μ_2), (λ_3, μ_3), (λ_4, μ_4). [M.T. II.]

10. Given a tetrahedron $ABCD$ and a quadric S, prove that the lines joining A, B, C, D to the poles A', B', C', D' of the opposite faces are generators of the same system of a quadric Φ. If E is any other point, show that the quadric Φ' determined in like manner from the tetrahedron $ABCE$ has with Φ a generator in common which passes through D', the pole of the plane ABC with respect to S. [M.T. II.]

11. Prove that the polar line l' of the line l joining the points whose parameters are (λ, μ) and (λ', μ') of the quadric $(\lambda\mu, \lambda, \mu, 1)$ is the join of the points whose parameters are (λ', μ) and (λ, μ'). Prove further that, if the line joining the points (α, β) and (α', β') meets both the lines l, l', then the cross-ratio

$$(\alpha, \alpha', \lambda, \lambda') = (\beta, \beta', \mu, \mu') = -1.$$ [M.T. II.]

12. Deduce from Example 11 that the transversals of two lines l_1, l_2 and their polar lines with respect to a quadric are, in general, themselves polar lines with respect to the quadric. [M.T. II.]

13. P_1, P_2 are two fixed points and P is a variable point such that each of the lines PP_1, PP_2 meets the polar line of the other with respect to a quadric S. Prove that the locus of P is the quadric which passes through P_1, P_2 and the sections of S by the polar planes of P_1 and P_2. [M.T. II.]

14. A skew hexagon is formed by the six lines a, b, c, d, e, f (in this order); it may be assumed that each side meets only those adjacent to it. From a point O, whose position is general with respect to the hexagon, transversals are drawn to opposite sides of the hexagon meeting the sides in A, B, C, D, E, F respectively. Prove that the six sides of the hexagon $ABCDEF$ lie on a quadric.

If a, b, c, d, e, f meet this quadric again in the points A', B', C', D', E', F' respectively, prove that the six sides of the hexagon $A'B'C'D'E'F'$ lie on the quadric, and that the lines $A'D', B'E', C'F'$ are concurrent. [M.T. II.]

15. g_1, g_2, g_3 are three generators of one system on a quadric, and l_1, l_2, l_3 are three generators of the opposite system. The lines g_i, l_j meet in P_{ij}. Prove that the lines $P_{12}P_{31}, P_{13}P_{21}, P_{22}P_{33}$ meet in a point, and that the planes $P_{11}P_{22}P_{33}, P_{12}P_{23}P_{31}, P_{13}P_{21}P_{32}$ meet in a line l whose polar line lies in the three planes $P_{11}P_{23}P_{32}, P_{13}P_{22}P_{31}, P_{12}P_{21}P_{33}$. [M.T. II.]

16. An involution is set up between the generators of one system on a quadric, in which the double lines are p and q. A is a point in general position, and the transversal from A to p and q is a line l whose polar line is l'. Prove that the transversals from A to the pairs of generators which correspond in the involution all lie in the plane containing A and l'. [M.T. II.]

17. Prove that, if a $(1, 1)$ algebraic correspondence is set up between the lines of complementary reguli, the locus of the intersection of corresponding lines is a conic.

Four coplanar points are taken on a non-singular quadric Q. Prove that the sixteen intersections of the eight generators through them lie four in each of four planes, and that the tetrahedron formed by these planes is self-polar with respect to Q. [M.T. II.]

18. Prove that the tangent planes at four points of a generator of a quadric form a pencil of planes whose cross-ratio is equal to that of the range of four points.

Three generators a, b, c of the same system of a quadric meet a generator d of the opposite system in A', B', C' respectively, and any plane meets a, b, c, d in A, B, C, D respectively. If E is any other point of the section of the quadric by this plane, prove that

$$E(A, B, C, D) = (A', B', C', D). \qquad \text{[L.]}$$

19. Prove that four lines in space have in general just two common transversals.

Four skew lines a, b, a', b' have just two common transversals g, h. The transversal from a point P to a, b determines with the transversal from P to a', b' a plane π. Prove that, when P describes a line l which meets both g and h but none of a, b, a', b', the plane π passes through a fixed line l', and show that the lines l, l' are mutually related. [L.]

20. Two skew lines l, m meet each of two generators a, b of a quadric S, but are not themselves generators of S; and l', m' are the polar lines of l, m with respect to S. Prove that, if l, m are not conjugate with respect to S, then l, m, l', m' belong to a regulus.

The plane α of a triangle PQR meets a, b at points D, E, and the polar lines, with respect to S, of the transversals from P, Q, R to a, b meet α at P', Q', R'. Prove that, when DE is not a generator of S, the triangles PQR, $P'Q'R'$ are in perspective from the intersection of α with the polar line of DE with respect to S. [L.]

21. Two fixed points A, B are conjugate with respect to a quadric S upon which neither lies. The lines joining A and B to a variable point P of S are met again by S at Q and R. Prove that the point of intersection of AR and BQ lies on S, and that QR meets the polar line t of AB with respect to S.

Show that, when P describes a given generator g of S, skew to AB, the points Q and R describe generators of S; and that the locus of QR is a quadric through AB, t and four generators of S, two of which remain fixed when g varies in one system on S. [L.]

22. Prove that two polar triangles with respect to a conic Σ are in perspective; and hence show that the tangents at three points A, B, C of Σ meet the lines BC, CA, AB respectively at three points in line.

$ABCD$ is a tetrahedron inscribed in a quadric S. Prove that the tangent planes to S at the vertices A, B, C, D meet the opposite faces in four lines of a regulus, provided that no two opposite edges of the tetrahedron are conjugate with respect to S. [L.]

23. Three skew lines a, b, c are met by three skew lines l_1, l_2, l_3 in sets of points (A_1, A_2, A_3), (B_1, B_2, B_3), (C_1, C_2, C_3) respectively. Prove that the triangles $A_1B_2C_3$, $B_3C_1A_2$, $C_2A_3B_1$ are in perspective in pairs with the same axis p; that the triangles $A_1B_3C_2$, $C_3A_2B_1$, $B_2C_1A_3$ are in perspective in pairs with the same axis q; and that the centres of the last three perspectives lie upon p.

Show also that p and q are polar lines with respect to the quadric abc. [L.; M.A.]

24. Show that the range in which a variable generator of one system on a quadric cuts any conic on the quadric is projective with the range in which it cuts any generator of the opposite system.

A plane π_0 meets four generators $g_r (r = 1, 2, 3, 4)$ of a quadric in their intersections $(g_r g'_r)$ with four generators of the opposite system, and the intersections $(g_r g'_{r+1})$ lie in a second plane π_1 (where for all n we define g'_{n+4} as g'_n). Prove that each set of generators forms a harmonic set.

Prove that the intersections $(g_r g'_{r+2})$ and $(g_r g'_{r+3})$ lie in two planes π_2, π_3 coaxial with [i.e. in a pencil with] π_0 and π_1, and that these four planes form a harmonic set. [L.]

CHAPTER IV

LINE GEOMETRY

1. Preliminary remarks. The reader is familiar with the geometry of points and lines in a plane and with the geometry of points and planes in space. We now proceed to consider the geometry of lines in space. Our first problem is to show how a system of coordinates can be attached to the lines, so that a line is uniquely determined when its coordinates are given and conversely.

2. The numbers l, m, n, l', m', n'. Let

$$P_1(x_1, y_1, z_1, t_1), \quad P_2(x_2, y_2, z_2, t_2)$$

be two distinct points. We define from their coordinates the six numbers

$$l = x_1 t_2 - x_2 t_1, \quad m = y_1 t_2 - y_2 t_1, \quad n = z_1 t_2 - z_2 t_1,$$

$$l' = y_1 z_2 - y_2 z_1, \quad m' = z_1 x_2 - z_2 x_1, \quad n' = x_1 y_2 - x_2 y_1.$$

In some text-books the definitions of l, m, n have signs opposite to those given here. The necessary adjustments are small, but it is usually desirable to state precisely the convention that is used.

We prove first that the ratios of these six numbers are unaltered if the points P_1, P_2 are replaced by *any* two points P_1', P_2' of the line p joining them. In fact, if

$$\mathbf{P}_1' = \lambda \mathbf{P}_1 + \mu \mathbf{P}_2, \quad \mathbf{P}_2' = \nu \mathbf{P}_1 + \rho \mathbf{P}_2,$$

then it is easy to show that

$$x_1' t_2' - x_2' t_1' = (\lambda\rho - \mu\nu)(x_1 t_2 - x_2 t_1),$$

$$y_1' z_2' - y_2' z_1' = (\lambda\rho - \mu\nu)(y_1 z_2 - y_2 z_1)$$

and the result is immediate.

Hence *the ratios of the six numbers l, m, n, l', m', n' are uniquely determined by the line p.*

Theorem-examples. 1. The coordinates of the points in which the line P_1P_2 meets the faces of the tetrahedron of reference can be expressed in the forms

$$\begin{array}{l}(\ 0,\ -n',\ m',\ -l\ ),\\(\ n',\ 0,\ -l',\ -m),\\(-m',\ l',\ 0,\ -n),\\(\ l,\ m,\ n,\ 0).\end{array}$$

[Consider $x_2\mathbf{P}_1 - x_1\mathbf{P}_2$, etc.]

2. The equations of the planes joining the line P_1P_2 to the vertices of the tetrahedron of reference can be expressed in the forms

$$\begin{aligned} -ny+mz-l't &= 0,\\ nx-lz-m't &= 0,\\ -mx+ly-n't &= 0,\\ l'x+m'y+n'z &= 0.\end{aligned}$$

3. *The numbers l, m, n, l', m', n' are connected by the fundamental relation*

$$ll'+mm'+nn' = 0.$$

[See also §4 below.]

4. The ratios of the six numbers l, m, n, l', m', n' are not proportional for distinct lines.

[Two distinct lines must meet at least two of the faces of the tetrahedron of reference in distinct points.]

5. The numbers l, m, n, l', m', n' define a line, and define it uniquely, when they satisfy the relation $ll'+mm'+nn' = 0$.

[Use Th.-ex. 1 above.]

3. The coordinates of a line. The numbers l, m, n, l', m', n' are suitable as coordinates for a line, because

(i) if the line is given, the ratios of the numbers are determined;

(ii) if the ratios are given, then the line is determined uniquely.

We therefore call the numbers the *coordinates of the line*, and we refer to them as *line-coordinates*. We shall also speak of 'the line (l, m, n, l', m', n')'.

Note that *the six coordinates cannot be taken at random, but they must satisfy the relation*

$$ll'+mm'+nn' = 0.$$

4. An algebraic theorem and some important deductions. Suppose that one line is defined by two points (x_1, y_1, z_1, t_1), (x_2, y_2, z_2, t_2)

and another by two points (X_1, Y_1, Z_1, T_1), (X_2, Y_2, Z_2, T_2). The coordinates of the lines are (l, m, n, l', m', n') and (L, M, N, L', M', N'), defined as in § 2. We leave to the reader the proof of the algebraic theorem that

$$\begin{vmatrix} x_1 & y_1 & z_1 & t_1 \\ x_2 & y_2 & z_2 & t_2 \\ X_1 & Y_1 & Z_1 & T_1 \\ X_2 & Y_2 & Z_2 & T_2 \end{vmatrix} = lL' + l'L + mM' + m'M + nN' + n'N.$$

[The reader familiar with the Laplace expansion of determinants will solve the problem mentally. It can also be established quite easily by expanding in terms of the first row and then expanding the third-order determinants which arise.]

We deduce the following results:

(i) $ll' + mm' + nn' = 0$.

Put $(x_1, y_1, z_1, t_1) \equiv (X_1, Y_1, Z_1, T_1)$ and $(x_2, y_2, z_2, t_2) \equiv (X_2, Y_2, Z_2, T_2)$. Then $(l, m, n, l', m', n') \equiv (L, M, N, L', M', N')$. Also the determinant vanishes, since two rows (two pairs of rows, in fact) are equal. Hence

$$ll' + mm' + nn' = 0.$$

(ii) *If two lines whose coordinates are*

$$(l, m, n, l', m', n'), \quad (L, M, N, L', M', N')$$

intersect, then

$$lL' + l'L + mM' + m'M + nN' + n'N = 0.$$

If they meet, then the four points named above are coplanar, and so the determinant vanishes. The result follows.

(iii) *If* $lL' + l'L + mM' + m'M + nN' + n'N = 0$, *then the two lines*

$$(l, m, n, l', m', n'), \quad (L, M, N, L', M', N')$$

intersect.

Since the expression vanishes, the determinant also vanishes. Hence the four points are coplanar and so the two lines intersect.

Notation. It is often convenient to denote the coordinates of two lines by the suffix notation

$$(l_1, m_1, n_1, l_1', m_1', n_1'), \quad (l_2, m_2, n_2, l_2', m_2', n_2').$$

We then denote the expression

$$l_1 l_2' + l_2 l_1' + m_1 m_2' + m_2 m_1' + n_1 n_2' + n_2 n_1'$$

by the symbol ϖ_{12}, which is called the *mutual invariant* of the two lines. Its vanishing is the condition that they should intersect.

Note that, in similar notation,

$$\varpi_{11} \equiv 2(l_1 l_1' + m_1 m_1' + n_1 n_1') = 0.$$

5. The line as the intersection of two planes. A line may be defined as the intersection of two planes

$$a_1 x + b_1 y + c_1 z + d_1 t = 0,$$
$$a_2 x + b_2 y + c_2 z + d_2 t = 0$$

and it is important to be able to find its line-coordinates in terms of the constants in these equations. To help in our investigation we write (without prejudice)

$$\lambda = a_1 d_2 - a_2 d_1, \quad \mu = b_1 d_2 - b_2 d_1, \quad \nu = c_1 d_2 - c_2 d_1,$$
$$\lambda' = b_1 c_2 - b_2 c_1, \quad \mu' = c_1 a_2 - c_2 a_1, \quad \nu' = a_1 b_2 - a_2 b_1.$$

Then the equation of the plane joining the line to the vertex X of the tetrahedron of reference is (on eliminating x)

$$-\nu' y + \mu' z - \lambda t = 0.$$

But (§ 2, Theorem-example 2) the equation of this plane, expressed in terms of (l, m, n, l', m', n'), is also

$$-ny + mz - l't = 0,$$

so that

$$\frac{n}{\nu'} = \frac{m}{\mu'} = \frac{l'}{\lambda}.$$

Proceeding similarly, we obtain the set of ratios

$$\frac{l}{\lambda'} = \frac{m}{\mu'} = \frac{n}{\nu'} = \frac{l'}{\lambda} = \frac{m'}{\mu} = \frac{n'}{\nu}.$$

Corollary. The line-coordinates of the line of intersection of the planes whose coordinates are (a_1, b_1, c_1, d_1), (a_2, b_2, c_2, d_2) are given by the relations

$$l = b_1 c_2 - b_2 c_1, \quad m = c_1 a_2 - c_2 a_1, \quad n = a_1 b_2 - a_2 b_1,$$
$$l' = a_1 d_2 - a_2 d_1, \quad m' = b_1 d_2 - b_2 d_1, \quad n' = c_1 d_2 - c_2 d_1.$$

6. To find the coordinates of the point in which a given line meets a given plane. Let the line-coordinates of the line be (l, m, n, l', m', n') and the equation of the given plane

$$ax+by+cz+dt = 0.$$

If $P_1(x_1, y_1, z_1, t_1)$, $P_2(x_2, y_2, z_2, t_2)$ are two points of the line, then the point P where the line meets the plane has coordinates

$$(\lambda x_1+\mu x_2,\ \lambda y_1+\mu y_2,\ \lambda z_1+\mu z_2,\ \lambda t_1+\mu t_2),$$

where

$$a(\lambda x_1+\mu x_2)+b(\lambda y_1+\mu y_2)+c(\lambda z_1+\mu z_2)+d(\lambda t_1+\mu t_2) = 0.$$

Hence the symbol $\mathbf{P}$ of the point is given by the equation

$$\mathbf{P} = (ax_2+by_2+cz_2+dt_2)\,\mathbf{P}_1-(ax_1+by_1+cz_1+dt_1)\,\mathbf{P}_2.$$

If we denote $\mathbf{P}$ by (ξ, η, ζ, τ), then

$$\begin{aligned} \xi &= \quad\quad\; bn'-cm'+dl, \\ \eta &= -an' \quad\quad +cl' \;+dm, \\ \zeta &= \;\; am'-bl' \quad\quad +dn, \\ \tau &= -al \;\; -bm-cn. \end{aligned}$$

COROLLARY. If the line lies wholly in the plane, then the four expressions for ξ, η, ζ, τ all vanish.

Theorem-example. The plane joining the line (l, m, n, l', m', n') to the point (A, B, C, D) is given by $\alpha x+\beta y+\gamma z+\delta t = 0$, where

$$\begin{aligned} \alpha &= \quad\quad\; Bn-Cm+Dl', \\ \beta &= -An \quad\quad +Cl \;+Dm', \\ \gamma &= \;\; Am-Bl \quad\quad +Dn', \\ \delta &= -Al' \;\; -Bm'-Cn'. \end{aligned}$$

7. The linear complex. Just as the linear equation

$$lx+my+nz = 0$$

is important in plane geometry and the linear equation

$$lx+my+nz+pt = 0$$

in the geometry of space of three dimensions, so we consider in line-geometry the system of lines whose line-coordinates satisfy a linear relation, which we write in the form

$$\Gamma \equiv a'l+al'+b'm+bm'+c'n+cn' = 0.$$

Such a system of lines is said to form a *linear complex*.

We prove certain fundamental results:

(i) *The lines of the complex* Γ *which pass through a general fixed point lie in a fixed plane.* (See also the Theorem-examples below.)

Let (ξ, η, ζ, τ) be the fixed point, and take (x, y, z, t) as an arbitrary point of a line through it belonging to the complex. Then, by the definitions of § 2,

$$l = \xi t - \tau x, \quad l' = \eta z - \zeta y, \quad \text{etc.},$$

so that

$$a'(\xi t - \tau x) + a(\eta z - \zeta y) + \text{etc.} = 0.$$

The points of such lines thus all lie in the plane

$$(-c\eta + b\zeta - a'\tau)x + (-a\zeta + c\xi - b'\tau)y + (-b\xi + a\eta - c'\tau)z + (a'\xi + b'\eta + c'\zeta)t = 0.$$

(ii) *The lines of the complex* Γ *which lie in a general fixed plane pass through a fixed point.* (See also the Theorem-examples below.)

Let the fixed plane have *coordinates* (A, B, C, D), and take (u, v, w, p) as an arbitrary plane through a line lying in the given plane and belonging to the complex. Then, by § 5,

$$l = Bw - Cv, \quad l' = Ap - Du, \quad \text{etc.},$$

so that

$$a'(Bw - Cv) + a(Ap - Du) + \text{etc.} = 0,$$

or

$$(-c'B + b'C - aD)u + (-a'C + c'A - bD)v + (-b'A + a'B - cD)w + (aA + bB + cC)p = 0.$$

The plane

$$ux + vy + wz + pt = 0$$

therefore passes through the fixed point (x_0, y_0, z_0, t_0) whose coordinates are given by the relations

$$\begin{aligned} x_0 &= -c'B + b'C - aD, \\ y_0 &= c'A - a'C - bD, \\ z_0 &= -b'A + a'B - cD, \\ t_0 &= aA + bB + cC. \end{aligned}$$

DEFINITION. A point and a plane related as in (i) and (ii) may be said to be *pole* and *polar* with respect to the complex.

(iii) *The* SPECIAL *linear complex.*

It may happen that a, b, c, a', b', c' are connected by the relation

$$aa'+bb'+cc' = 0.$$

In that case, the six numbers are the coordinates of a line λ, and each line of the complex meets λ. The complex is then said to be *special*, and λ is called the *axis* of the complex.

Conversely, all the lines which meet a given line λ belong to a linear complex, as we see at once by writing down the condition given in § 4.

Note that *the line λ itself belongs to the linear complex.*

Theorem-examples. 1. The polar plane of a point on the axis of a special complex is indeterminate.

[The reader may find it helpful to consider the special complex $l' = 0$ and the point $(1, 0, 0, 0)$ on the axis. The equation given in (i) vanishes identically.]

2. The pole of a plane through the axis of a special complex is indeterminate.

3. The polar planes of the points of an arbitrary given line pass through a fixed line.

[Take the points of the arbitrary given line in the form $\lambda_1 \mathbf{P}_1 + \lambda_2 \mathbf{P}_2$.]

4. The poles of the planes through an arbitrary given line lie on a fixed line.

8. The linear congruence. The lines whose line-coordinates satisfy two linear equations

$$\Gamma_i \equiv a_i' l + a_i l' + b_i' m + b_i m' + c_i' n + c_i n' = 0 \quad (i = 1, 2)$$

are said to form a *linear congruence*. Thus a linear congruence is a system of lines common to two linear complexes. The fundamental theorem, which we now prove, is the following:

The lines of a linear congruence all meet each of two (possibly 'coincident') fixed lines. These lines are called the *directrices* of the congruence.

The lines which satisfy the two equations $\Gamma_1 = 0$, $\Gamma_2 = 0$ also satisfy the equation

$$\lambda_1 \Gamma_1 + \lambda_2 \Gamma_2 = 0,$$

or

$$(\lambda_1 a_1' + \lambda_2 a_2') l + (\lambda_1 a_1 + \lambda_2 a_2) l' + \text{etc.} = 0. \tag{1}$$

We write $\varpi_{ij} \equiv a_i a'_j + a_j a'_i + b_i b'_j + b_j b'_i + c_i c'_j + c_j c'_i$

and choose the ratio $\lambda_1 : \lambda_2$ so that

$$(\lambda_1 a_1 + \lambda_2 a_2)(\lambda_1 a'_1 + \lambda_2 a'_2) + \text{etc.} = 0,$$

or

$$\varpi_{11}\lambda_1^2 + 2\varpi_{12}\lambda_1\lambda_2 + \varpi_{22}\lambda_2^2 = 0.$$

The numbers $\lambda_1 a_1 + \lambda_2 a_2, \quad \ldots, \quad \lambda_1 a'_1 + \lambda_2 a'_2, \quad \ldots$

are then the coordinates of a line, and this line, by equation (1), meets every line of the congruence. But the equation for $\lambda_1 : \lambda_2$ is quadratic, and has therefore two (possibly 'coincident') solutions, as required.

In the particular case when Γ_1, Γ_2 are special linear complexes whose defining lines intersect, the equation for $\lambda_1 : \lambda_2$ vanishes identically. The reader should interpret this fact geometrically.

9. The regulus. We have already met the regulus as a system of lines generating a quadric surface. We now prove that *the lines whose line-coordinates satisfy three linear equations*

$$\Gamma_i \equiv a'_i l + a_i l' + b'_i m + b_i m' + c'_i n + c_i n' = 0 \quad (i = 1, 2, 3)$$

belong (*in the general case*) *to a regulus.*

The lines which satisfy the three equations $\Gamma_1 = 0$, $\Gamma_2 = 0$, $\Gamma_3 = 0$ also satisfy the equation $\lambda_1 \Gamma_1 + \lambda_2 \Gamma_2 + \lambda_3 \Gamma_3 = 0$, or

$$(\lambda_1 a'_1 + \lambda_2 a'_2 + \lambda_3 a'_3) l + (\lambda_1 a_1 + \lambda_2 a_2 + \lambda_3 a_3) l' + \text{etc.} = 0. \qquad (1)$$

We use the notation ϖ_{ij} defined in § 8, and choose the ratios $\lambda_1 : \lambda_2 : \lambda_3$ so that

$$(\lambda_1 a_1 + \lambda_2 a_2 + \lambda_3 a_3)(\lambda_1 a'_1 + \lambda_2 a'_2 + \lambda_3 a'_3) + \text{etc.} = 0,$$

or $\varpi_{11}\lambda_1^2 + \varpi_{22}\lambda_2^2 + \varpi_{33}\lambda_3^2 + 2\varpi_{23}\lambda_2\lambda_3 + 2\varpi_{31}\lambda_3\lambda_1 + 2\varpi_{12}\lambda_1\lambda_2 = 0.$

This can be done in an infinity of ways, and the numbers

$$\lambda_1 a_1 + \lambda_2 a_2 + \lambda_3 a_3, \quad \ldots, \quad \lambda_1 a'_1 + \lambda_2 a'_2 + \lambda_3 a'_3, \quad \ldots$$

are then the coordinates of a line which, by equation (1), meets every line of the system. We have therefore defined two systems of lines with the property that *every* line of each system meets *every* line of the other. They are therefore (in the general case) the two systems of generators of a quadric.

LEMMA. If, by analogy with the similar case for a point, we use the notation $\mathbf{p}$ to denote the *symbol* of a line $p(l, m, n, l', m', n')$, then the existence of a syzygy

$$a\mathbf{p}_1 + b\mathbf{p}_2 + c\mathbf{p}_3 + d\mathbf{p}_4 \equiv 0$$

indicates that the four lines p_1, p_2, p_3, p_4 belong to a regulus.

10. The quadric whose generators of one system are the lines common to three linear complexes. Let a line (l, m, n, l', m', n') belong to each of the linear complexes

$$\Gamma_i \equiv a_i' l + b_i' m + c_i' n + a_i l' + b_i m' + c_i n' = 0 \quad (i = 1, 2, 3).$$

Then (§ 2, Theorem-example 2), that line lies in each of the planes

$$\begin{aligned} zm - yn - tl' &= 0, \\ -zl + xn - tm' &= 0, \\ yl - xm - tn' &= 0, \\ xl' + ym' + zn' &= 0, \end{aligned}$$

and so we have seven linear equations connecting the six homogeneous coordinates l, m, n, l', m', n'. We can eliminate these coordinates determinantally by choosing six of the equations which are *linearly* independent. Let us, for convenience, choose the first six. We then obtain the equation

$$\begin{vmatrix} a_1' & b_1' & c_1' & a_1 & b_1 & c_1 \\ a_2' & b_2' & c_2' & a_2 & b_2 & c_2 \\ a_3' & b_3' & c_3' & a_3 & b_3 & c_3 \\ 0 & z & -y & -t & 0 & 0 \\ -z & 0 & x & 0 & -t & 0 \\ y & -x & 0 & 0 & 0 & -t \end{vmatrix} = 0$$

connecting x, y, z, t.

This equation, on expansion, is cubic in x, y, z, t, but the left-hand side contains the irrelevant factor $t = 0$ since the determinant vanishes when t is zero. (If the first, second, or third of the equations in x, y, z, t had been omitted, the corresponding irrelevant factor would have been x, y, z respectively.) On removing that factor, we obtain the equation of the quadric generated by the regulus.

Note that the six equations chosen are *linearly* independent, though the three involving x, y, z, t are equivalent to two only in virtue of the *quadratic* relation $ll'+mm'+nn' = 0$. The last three rows of the determinant are linearly independent.

It is a simple matter to derive from this result the equation of the quadric of which three given skew lines are generators.

11. The lines common to four linear complexes. The lines whose line-coordinates satisfy the four equations

$$\Gamma_i \equiv a_i' l + a_i l' + b_i' m + b_i m' + c_i' n + c_i n' = 0 \quad (i = 1, 2, 3, 4)$$

are in general *two* in number (possibly 'coincident'). Their line-coordinates can be found—in theory, at any rate—by solving the five equations

$$\Gamma_1 = 0, \quad \Gamma_2 = 0, \quad \Gamma_3 = 0, \quad \Gamma_4 = 0,$$

$$ll' + mm' + nn' = 0.$$

12. Polar lines with respect to a quadric. Let

$$S \equiv ax^2 + by^2 + cz^2 + dt^2 = 0$$

be a quadric with respect to which the tetrahedron of reference is self-polar, and let $\lambda(l, m, n, l', m', n')$ be a line in arbitrary position. We find the line-coordinates (L, M, N, L', M', N') of the polar line of λ with respect to S.

If $P_1(x_1, y_1, z_1, t_1)$, $P_2(x_2, y_2, z_2, t_2)$ are two points of λ, then the polar line of λ is the intersection of the planes

$$ax_1 x + by_1 y + cz_1 z + dt_1 t = 0,$$

$$ax_2 x + by_2 y + cz_2 z + dt_2 t = 0,$$

and so (§ 5) $\quad L = by_1 . cz_2 - by_2 . cz_1 = bcl',$

$$L' = ax_1 . dt_2 - ax_2 . dt_1 = adl, \text{ etc.}$$

Hence *the polar line of $\lambda(l, m, n, l', m', n')$ with respect to the quadric $ax^2 + by^2 + cz^2 + dt^2 = 0$ is the line*

$$\mu(bcl', cam', abn', adl, bdm, cdn).$$

We can use this result to find the generators of the quadric; for a generator is its own polar line, so that the line-coordinates of a generator satisfy the equations

$$\frac{l}{bcl'} = \frac{m}{cam'} = \frac{n}{abn'} = \frac{l'}{adl} = \frac{m'}{bdm} = \frac{n'}{cdn}.$$

If we denote each of these ratios by σ, then

$$l = bcl'\sigma, \quad l' = adl\sigma$$

and thus $$abcd\sigma^2 = 1.$$

It follows that $$\sigma = \pm \frac{1}{\sqrt{(abcd)}}.$$

The coordinates of a generator of one system can be taken in the form

$$(l,\ m,\ n,\quad adl\sigma,\quad bdm\sigma,\quad cdn\sigma)$$

and the coordinates of a generator of the other system can be taken in the form

$$(l,\ m,\ n,\quad -adl\sigma,\quad -bdm\sigma,\quad -cdn\sigma),$$

where, in each case, l, m, n are arbitrary numbers subject to the condition

$$al^2 + bm^2 + cn^2 = 0.$$

The coordinates of these generators can, of course, be expressed in many apparently different forms. The form which we have given seems to be convenient.

Theorem-examples. 1. Two generators of the same system do not intersect. [Use the condition of § 4 (ii).]

2. Two generators of opposite systems intersect.

13. The quadratic complex. There are many complexes of lines other than the linear complex considered in § 7, but we do not propose to go into their properties in much detail. We conclude this discussion with a sketch of some typical results.

The lines whose line-coordinates satisfy a quadratic equation are said to form a *quadratic* complex. For example, the equation

$$all' + bmm' + cnn' = 0,$$

where a, b, c are not all equal, defines a (particular) quadratic complex.

Theorem-examples. 1. The lines which belong to a quadratic complex and which pass through a fixed point generate (in the general case) a quadric cone.

2. The lines which belong to a quadratic complex and which lie in a fixed plane are (in the general case) the tangents of a conic.

[The proofs of 1 and 2 are very similar to those given in the corresponding cases for a linear complex.]

14. The tetrahedral complex. One particular example of a quadratic complex can be treated simply. We first require a lemma.

LEMMA. *To find an expression for the cross-ratio of the four points in which the line* (l, m, n, l', m', n') *meets the faces of the tetrahedron of reference.*

The four points are (§ 2, Theorem-example 1)

$$A(0, -n', m', -l), \quad B(n', 0, -l', -m), \quad C(-m', l', 0, -n),$$
$$D(l, m, n, 0).$$

Thus

$$l\mathbf{B} \equiv m\mathbf{A} + n'\mathbf{D},$$
$$l\mathbf{C} \equiv n\mathbf{A} - m'\mathbf{D},$$

so that

$$(D, A, B, C) = (0, \infty, m/n', -n/m')$$
$$= -\frac{mm'}{nn'}.$$

The six values of the cross-ratio, for varying orders of the letters A, B, C, D, are in fact the six ratios of pairs of numbers ll', mm', nn', with the sign changed.

DEFINITION. If the quadratic relation between the line-coordinates is of the form

$$all' + bmm' + cnn' = 0,$$

the lines are said to belong to a *tetrahedral* complex.

Theorem-examples. 1. The lines of a tetrahedral complex cut the faces of the tetrahedron of reference in four points whose cross-ratio is constant.

[Solve the equations $all'+bmm'+cnn' = 0$, $ll'+mm'+nn' = 0$ for the ratios $ll':mm':nn'$.]

2. The lines which cut the faces of the tetrahedron of reference in four points whose cross-ratio is constant all belong to a tetrahedral complex.

3. The lines of the tetrahedral complex $all'+bmm'+cnn' = 0$ which pass through the point $P(\xi, \eta, \zeta, \tau)$ generate the quadric cone

$$a(\tau x-\xi t)(\zeta y-\eta z)+b(\tau y-\eta t)(\xi z-\zeta x)+c(\tau z-\zeta t)(\eta x-\xi y) = 0.$$

If P lies in the plane XYZ, then the cone degenerates into two planes, one of which is the plane XYZ and the other of which passes through the vertex T.

[For the general quadratic complex, the locus of points (such as P) for which the cone degenerates into two planes is a quartic surface called the **singular surface** of the complex. For the tetrahedral complex, the singular surface is itself degenerate, consisting of the four faces of the tetrahedron.]

ILLUSTRATION. *The poles, with respect to a quadric, of the faces YZT, ZXT, XYT, XYZ of a tetrahedron $XYZT$ are X', Y', Z', T' respectively. To prove that the lines XX', YY', ZZ', TT' are generators of one system on a quadric.*

Take $XYZT$ as the tetrahedron of reference and suppose that the tangential equation of the quadric is

$$al^2+bm^2+cn^2+dp^2+2fmn+2gnl+2hlm \\ +2ulp+2vmp+2wnp = 0.$$

Then the equation of X', the pole of the plane $(1, 0, 0, 0)$, is

$$al+hm+gn+up = 0,$$

so that X' is the point (a, h, g, u). The line-coordinates of XX' are thus

$$p_1(u, 0, 0, 0, -g, h).$$

In the same way, the line-coordinates of YY', ZZ', TT' are

$$\begin{array}{l} p_2(\ 0,\ \ v,\ \ 0,\ \ f,\ \ 0, -h), \\ p_3(\ 0,\ \ 0,\ \ w, -f,\ \ g,\ \ 0), \\ p_4(-u, -v, -w,\ \ 0,\ \ 0,\ \ 0). \end{array}$$

But

$$\mathbf{p}_1+\mathbf{p}_2+\mathbf{p}_3+\mathbf{p}_4 \equiv 0,$$

and so (§ 9, Lemma, p. 73) the four lines belong to a regulus.

EXAMPLES IV

1. Find the line-coordinates of the line joining the points $(1, 3, 5, 7)$, $(3, 1, -1, -3)$ and of the line joining the points $(8, 3, 2, 1)$, $(2, 1, 0, -1)$. Verify from the condition of § 4 (iii) that the lines intersect.

2. Find the line-coordinates of the line of intersection of the planes

$$x+y+z+t = 0, \quad 3x+4y+2z+t = 0.$$

Use the formulae of § 2, Theorem-example 1, to find the coordinates of the points in which the line meets the faces of the tetrahedron of reference, and verify the results by direct calculation.

3. Find the line-coordinates of the line joining the points $(1, 2, 3, -1)$, $(4, 3, 3, -4)$.

Use the formula of § 6, Theorem-example, to prove that the equation of the plane joining this line to the point $(1, 0, 2, -1)$ is $x+t = 0$.

4. Prove that the linear complex

$$l+2m+2n+2l'+m'-2n' = 0$$

is special, and give the line-coordinates of the axis.

Find the equation of the plane containing those lines of the complex which pass through the point $(1, 1, 1, 1)$, and verify that this plane contains the axis of the complex.

5. Prove (do not merely verify) that the line-coordinates of the directrices of the congruence given by the equations

$$l+m+n+l'+m'+n' = 0, \quad l-2l' = 0$$

are $$(-1, 1, 1, 2, 1, 1), \quad (8, 2, 2, -1, 2, 2).$$

6. Prove that the line-coordinates of the polar line of the line

$$(l, m, n, l', m', n')$$

with respect to the quadric $xt-yz = 0$ are $(l', m, -n, l, m', -n')$.

MISCELLANEOUS EXAMPLES IV

1. Prove that the condition that two lines λ_i, λ_j in three dimensions should meet is

$$\varpi_{ij} \equiv l_i l'_j + l'_i l_j + m_i m'_j + m'_i m_j + n_i n'_j + n'_i n_j = 0,$$

where $(l_i, m_i, n_i, l'_i, m'_i, n'_i)$ are the coordinates of the line λ_i.

Show that the coordinates of the generators of one system of the quadric containing the non-intersecting lines λ_1, λ_2, λ_3 are given by

$$L = \alpha_1 l_1 + \alpha_2 l_2 + \alpha_3 l_3, \quad L' = \alpha_1 l'_1 + \alpha_2 l'_2 + \alpha_3 l'_3,$$

with similar expressions for M, M', N, N', where $\alpha_1, \alpha_2, \alpha_3$ satisfy the relation

$$\frac{\varpi_{23}}{\alpha_1}+\frac{\varpi_{31}}{\alpha_2}+\frac{\varpi_{12}}{\alpha_3}=0.$$

Deduce that the two transversals of the four non-intersecting lines $\lambda_1, \lambda_2, \lambda_3, \lambda_4$ coincide if

$$(\varpi_{23}\varpi_{14})^{\frac{1}{2}}+(\varpi_{31}\varpi_{24})^{\frac{1}{2}}+(\varpi_{12}\varpi_{34})^{\frac{1}{2}}=0. \qquad \text{[P.]}$$

2. Prove that the polar planes of any point P in regard to the quadrics

$$(a+\lambda)x^2+(b+\lambda)y^2+(c+\lambda)z^2+(d+\lambda)t^2=0,$$

where λ is variable, all pass through the same line p, and that, whatever point P is taken, the six coordinates of p satisfy the relation

$$(bc+ad)\,ll'+(ca+bd)\,mm'+(ab+cd)\,nn'=0.$$

Prove that if the lines p_1 and p_2, which arise in this way from two points P_1 and P_2, intersect, then the lines arising from all the points of the line P_1P_2 pass through their point of intersection. [P.]

3. Prove that the lines of a linear complex which meet a line p meet a second line p'. If p describes one system of generators of a quadric, show that p' generates another quadric. [P.]

4. Show that there is one linear complex containing five arbitrary lines. Denoting by $(l_i, m_i, n_i, l_i', m_i', n_i')$ the coordinates of a line a_i, prove that, if the linear complex determined by a_1, a_2, a_3, a_4, a_5 is special, i.e., if the five lines have a common transversal, then

$$\begin{vmatrix} 0 & \varpi_{12} & \varpi_{13} & \varpi_{14} & \varpi_{15} \\ \varpi_{21} & 0 & \varpi_{23} & \varpi_{24} & \varpi_{25} \\ \varpi_{31} & \varpi_{32} & 0 & \varpi_{34} & \varpi_{35} \\ \varpi_{41} & \varpi_{42} & \varpi_{43} & 0 & \varpi_{45} \\ \varpi_{51} & \varpi_{52} & \varpi_{53} & \varpi_{54} & 0 \end{vmatrix}=0,$$

where $\varpi_{ij}=l_il_j'+l_jl_i'+m_im_j'+m_jm_i'+n_in_j'+n_jn_i'$. [M.T. II.]

5. Prove that the polar line with respect to the quadric

$$ax^2+by^2+cz^2+dt^2=0$$

of the line whose coordinates are (l, m, n, l', m', n') has coordinates

$$(bcl', cam', abn', adl, bdm, cdn).$$

Prove also that the generators of the quadric have coordinates

$$(kl, km, kn, al, bm, cn),$$

where $al^2+bm^2+cn^2=0$, $k=\pm\sqrt{(abc/d)}$.

If three generators of one system of the quadric $ax^2+by^2+cz^2+dt^2=0$ are such that each meets the polar lines of the other two with respect to the quadric $x^2+y^2+z^2+t^2=0$, prove that

$$\frac{a}{k^2+a^2}+\frac{b}{k^2+b^2}+\frac{c}{k^2+c^2}=0. \qquad \text{[M.T. II.]}$$

6. Three linear complexes are defined by the equations

$$K_i \equiv a_i'l+a_il'+b_i'm+b_im'+c_i'n+c_in'=0.$$

Show that the condition for the line (l,m,n,l',m',n') to touch the quadric is

$$\begin{vmatrix} \varpi_{11} & \varpi_{12} & \varpi_{13} & K_1 \\ \varpi_{21} & \varpi_{22} & \varpi_{23} & K_2 \\ \varpi_{31} & \varpi_{32} & \varpi_{33} & K_3 \\ K_1 & K_2 & K_3 & 0 \end{vmatrix} = 0,$$

where $$2\varpi_{ij}=a_ia_j'+a_ja_i'+b_ib_j'+b_jb_i'+c_ic_j'+c_jc_i'. \qquad \text{[M.T. II.]}$$

7. Verify that the lines whose line-coordinates satisfy the relations

$$\frac{m}{a}=\frac{l+l'}{b}=\frac{m'}{c}$$

all meet two fixed generators of the quadric $xt=yz$. [M.T. II.]

8. The two linear complexes

$$2l+m+n+l'+m'+n'=0,$$

$$l-m-l'+m'-2n'=0$$

each contain two generators of each system on the quadric $xt-yz=0$. Prove that the cross-ratio of the four generators of one system is harmonic and that the cross-ratio of the four generators of the other system is equi-anharmonic. [M.T. II.]

9. Find in point coordinates the equation of the quadric generated by the lines common to the three complexes

$$l+m=0, \quad l'+m'=0, \quad l+n+l'+n'=0. \qquad \text{[M.T. II.]}$$

10. Prove that all the lines of a linear complex which meet a fixed line l meet another fixed line l'.

Five lines p_1, p_2, p_3, p_4, p_5 are in general position. The two transversals of p_2, p_3, p_4, p_5 are u_1, v_1, and pairs of lines u_2, v_2; u_3, v_3; u_4, v_4; u_5, v_5 are similarly defined. A general plane π meets $u_1, v_1, \ldots, u_5, v_5$ in points $U_1, V_1, \ldots, U_5, V_5$. Prove that the lines $U_1V_1, U_2V_2, \ldots, U_5V_5$ are concurrent. [P.]

11. Show that the polar line of the line (l, m, n, l', m', n') with regard to the quadric $x^2+y^2+z^2+t^2 = 0$ is the line (l', m', n', l, m, n). Deduce that the given line is a generator of the quadric *if, and only if,*

$$\frac{l}{l'} = \frac{m}{m'} = \frac{n}{n'} = \pm 1.$$

Prove that the polar lines with regard to the quadric of the lines of a given linear complex form a complex which is identical with the first *if, and only if,* one of the two systems of generators of the quadric belongs to the complex. [L.; advanced.]

12. A projectivity is established between the points of two non-intersecting lines p, q; P_1, P_2 are any two points of p, and Q_1, Q_2 the corresponding points of q. Prove that the lines which meet P_1Q_2, P_2Q_1 belong to a linear complex K which is independent of the choice of P_1, P_2, and that all the lines of K are thus obtained.

Show further that any general linear complex can be so defined in terms of two of its lines p, q. [P.]

CHAPTER V

THE TWISTED CUBIC

1. Definition and first properties. We have seen that a quadric surface can be obtained as the locus of the line of intersection of corresponding planes of two related pencils. We now consider the locus of the point of intersection of corresponding planes of three related pencils. This is in general a curve, called a *twisted cubic*. Clearly it lies on each of the three quadrics defined by two of the related pencils.

If $L_1 = 0, L_2 = 0$ are two planes of the first pencil, $M_1 = 0, M_2 = 0$ the corresponding planes of the second pencil and $N_1 = 0$, $N_2 = 0$ the corresponding planes of the third pencil, then arbitrary corresponding planes, one from each pencil, can be taken in the form [Compare M., pp. 30–1]

$$\varpi_1 \equiv L_1 + \theta L_2 = 0,$$
$$\varpi_2 \equiv M_1 + \theta M_2 = 0,$$
$$\varpi_3 \equiv N_1 + \theta N_2 = 0,$$

where θ is a parameter which varies from plane to plane.

By eliminating θ between pairs of these equations, we see that the point of intersection of the planes ϖ_1, ϖ_2, ϖ_3 lies on each of the quadrics

$$S_1 \equiv M_1 N_2 - M_2 N_1 = 0,$$
$$S_2 \equiv N_1 L_2 - N_2 L_1 = 0,$$
$$S_3 \equiv L_1 M_2 - L_2 M_1 = 0.$$

Now the quadrics S_2, S_3 have, by inspection, the line $L_1 = L_2 = 0$ as a common generator, and this line is not, in general, part of the locus, since it need not lie on S_1. The twisted cubic is therefore the curve of intersection of the quadrics S_2, S_3 *residual* to that line.

We are deliberately slurring over a number of points which ought to be mentioned in a full discussion, but over-elaboration of detail at the present stage can become somewhat tedious. The words 'in general' should be kept at the back of the mind throughout this introduction. The reader may find it helpful to glance from time to time at the particular case

$$x-\theta y = 0,$$
$$y-\theta z = 0,$$
$$z-\theta t = 0.$$

The quadrics are

$$S_1 \equiv z^2-yt = 0, \quad S_2 \equiv xt-yz = 0, \quad S_3 \equiv y^2-zx = 0.$$

Then the generator common to S_2, S_3 is the line $x = y = 0$, which does not lie on S_1.

It follows from what has been said that *a twisted cubic curve can be obtained as the residual intersection of two quadrics with a common generator*. Our work has led to three such pairs of quadrics; there are, in fact, infinitely many.

Another geometrical interpretation of the algebra may also be noted. First of all, we define a *star* of planes to be the system of planes passing through a fixed point. If $\alpha = 0$, $\beta = 0$, $\gamma = 0$ are the equations of three given planes belonging to one star, then the equation of a general plane of the star can be taken in the form

$$\lambda\alpha+\mu\beta+\nu\gamma = 0,$$

where λ, μ, ν are parameters which vary from plane to plane.

The reader may find it helpful to consider the special case in which α, β, γ are three faces of the tetrahedron of reference.

Suppose next that α', β', γ' are three planes of another star; then a general plane of that star can be taken in the form

$$\lambda'\alpha'+\mu'\beta'+\nu'\gamma' = 0.$$

We say that the stars are *related* if λ', μ', ν' are proportional to homogeneous linear functions of λ, μ, ν.

Now it is a theorem of plane geometry that in such a case the equations defining the relation can be reduced to the simple form

$$\lambda' = \lambda, \quad \mu' = \mu, \quad \nu' = \nu.$$

See, for example, J. A. Todd, *Projective and Analytical Geometry*, § 1·10.

It follows that the equations of corresponding planes of our related stars can be taken in the form

$$\lambda\alpha + \mu\beta + \nu\gamma = 0,$$
$$\lambda\alpha' + \mu\beta' + \nu\gamma' = 0.$$

Returning to the twisted cubic, we take the equations of two related stars, in which the planes L_1, L_2; M_1, M_2; N_1, N_2 correspond, to be

$$\pi_1 \equiv \lambda L_1 + \mu M_1 + \nu N_1 = 0,$$
$$\pi_2 \equiv \lambda L_2 + \mu M_2 + \nu N_2 = 0.$$

We prove that *corresponding planes of these stars intersect in chords of the twisted cubic:*

The planes π_1, π_2 meet in a line l, and we have to prove that two points of the twisted cubic lie on l. The cubic itself is, by definition, traced out by the point of intersection of the planes

$$\varpi_1 \equiv L_1 + \theta L_2 = 0,$$
$$\varpi_2 \equiv M_1 + \theta M_2 = 0,$$
$$\varpi_3 \equiv N_1 + \theta N_2 = 0.$$

Now ϖ_1 meets l in a point P, and ϖ_2 meets l in a point Q. If P and Q coincide, the coordinates of the point of coincidence satisfy the equations

$$\varpi_1 = \varpi_2 = 0.$$

But
$$\pi_1 + \theta\pi_2 \equiv \lambda\varpi_1 + \mu\varpi_2 + \nu\varpi_3,$$

and π_1, π_2 both vanish for points on l. Hence also

$$\varpi_3 = 0.$$

Since ϖ_1, ϖ_2, ϖ_3 all vanish, the point of coincidence lies on the twisted cubic.

Moreover as θ varies, P and Q trace related ranges on l, and these have two common corresponding points. That is, two points of the twisted cubic lie on l, which is therefore a chord.

2. The standard parametric form. The equations of the planes whose intersection defines the twisted cubic can be taken as

$$(a_1+\theta a_1')\,x+(b_1+\theta b_1')\,y+(c_1+\theta c_1')\,z+(d_1+\theta d_1')\,t=0,$$

$$(a_2+\theta a_2')\,x+(b_2+\theta b_2')\,y+(c_2+\theta c_2')\,z+(d_2+\theta d_2')\,t=0,$$

$$(a_3+\theta a_3')\,x+(b_3+\theta b_3')\,y+(c_3+\theta c_3')\,z+(d_3+\theta d_3')\,t=0.$$

To find the coordinates of their common point, we solve these equations in the usual way by 'Cramer's Rule'. It then appears that x, y, z, t are proportional to numbers which are (in general) cubic in the parameter θ, as we see on expanding the determinants used in the rule. We thus obtain solutions which we write in the form

$$\frac{x}{p_1\theta^3+q_1\theta^2+r_1\theta+s_1}=\frac{y}{p_2\theta^3+q_2\theta^2+r_2\theta+s_2}=\frac{z}{p_3\theta^3+q_3\theta^2+r_3\theta+s_3}=\frac{t}{p_4\theta^3+q_4\theta^2+r_4\theta+s_4}.$$

Hence *the coordinates of the points of a twisted cubic curve can be expressed as cubic polynomials in a parameter* θ.

The process can now be reversed, and these last equations can, in general, be solved to give θ^3, θ^2, θ, 1 as linear functions of x, y, z, t. We thus have expressions of the form

$$\frac{\theta^3}{l_1x+m_1y+n_1z+p_1t}=\frac{\theta^2}{l_2x+m_2y+n_2z+p_2t}=\frac{\theta}{l_3x+m_3y+n_3z+p_3t}=\frac{1}{l_4x+m_4y+n_4z+p_4t}.$$

[The use of p_1, p_2, p_3, p_4 in each of the last two sets of equations will not confuse. The double use of the symbols is for convenience of notation and the values are different in the two cases.]

Finally, we apply to the coordinate system the transformation, non-singular in the general case,

$$x' = l_1x+m_1y+n_1z+p_1t,$$
$$y' = l_2x+m_2y+n_2z+p_2t,$$
$$z' = l_3x+m_3y+n_3z+p_3t,$$
$$t' = l_4x+m_4y+n_4z+p_4t.$$

Dropping dashes, we obtain the simple form

$$\frac{x}{\theta^3} = \frac{y}{\theta^2} = \frac{z}{\theta} = \frac{t}{1},$$

so that *the coordinates of the points of a twisted cubic can be expressed parametrically in the form* $(\theta^3, \theta^2, \theta, 1)$.

The form $$\frac{x}{\theta^3} = \frac{y}{\theta^2} = \frac{z}{\theta} = \frac{t}{1}$$

is precisely the one obtained from the particular example quoted above (p. 83). This shows that, under suitable circumstances, the equations *can* be solved in the form stated, and also that there need not be any linear equation connecting x, y, z, t; in other words, it shows that the twisted cubic is a genuine *space* curve, not restricted to lie in a plane.

On the other hand, particular examples can easily be constructed where the general analysis breaks down. Thus consider the related pencils

$$x = \theta y,$$
$$x = \theta z,$$
$$z = \theta t.$$

Solving by Cramer's Rule (a heavy method for the particular case, but necessary to illustrate the theory), we have

$$\frac{x}{\theta^3} = \frac{y}{\theta^2} = \frac{z}{\theta^2} = \frac{t}{\theta}.$$

The curve is degenerate, consisting of two parts:

(i) The points for which $\theta \neq 0$ lie in the plane $y = z$, and describe the *conic* in which that plane is cut by the cone $y^2 = xt$;

(ii) The points for which $\theta = 0$ lie (as we see from the original equations) on the *straight line* $x = z = 0$.

The conic and the line together make up the complete locus.

3. Properties deduced from the parametric form. We now assume that the coordinates of the twisted cubic curve, Γ say, are taken in the form $(\theta^3, \theta^2, \theta, 1)$, and we proceed to establish some important properties.

(i) *The curve Γ meets an arbitrary plane in three points.*

The equation of an arbitrary plane is

$$lx+my+nz+pt = 0,$$

and the point $(\theta^3, \theta^2, \theta, 1)$ of Γ lies on it if

$$l\theta^3+m\theta^2+n\theta+p = 0.$$

This equation is cubic in θ and has three solutions (not necessarily distinct), corresponding to three points in which the curve meets the plane.

(ii) *The equation of the plane through the three points with parameters α, β, γ is*

$$x-y(\alpha+\beta+\gamma)+z(\beta\gamma+\gamma\alpha+\alpha\beta)-t\alpha\beta\gamma = 0.$$

This is an almost immediate consequence of (i), for the corresponding cubic equation in θ has roots α, β, γ.

(iii) *The equations of the chord joining the points with parameters α, β are*

$$x-y(\alpha+\beta)+z\alpha\beta = 0,$$

$$y-z(\alpha+\beta)+t\alpha\beta = 0.$$

Suppose that the parameter of the third point in which an arbitrary plane through the chord meets Γ is θ. Then, by (ii), the equation of that plane is

$$x-y(\alpha+\beta+\theta)+z(\beta\theta+\theta\alpha+\alpha\beta)-t\alpha\beta\theta = 0$$

or $$\{x-y(\alpha+\beta)+z\alpha\beta\}-\theta\{y-z(\alpha+\beta)+t\alpha\beta\} = 0,$$

and this plane, as θ varies, passes through the line whose equations are

$$x-y(\alpha+\beta)+z\alpha\beta = 0,$$

$$y-z(\alpha+\beta)+t\alpha\beta = 0.$$

4. Tangent line and osculating plane. In the equations of § 3 (iii) let $\beta = \alpha$. Then the equations of the line become

$$x - 2y\alpha + z\alpha^2 = 0,$$
$$y - 2z\alpha + t\alpha^2 = 0.$$

This line is called the *tangent* to the curve at the point whose parameter is α.

In the equation of § 3 (ii) let β, γ each be equal to α. Then the equation of the plane becomes

$$x - 3y\alpha + 3z\alpha^2 - t\alpha^3 = 0.$$

This plane is called the *osculating plane* to the curve at the point whose parameter is α.

Note. More generally, suppose that the coordinates of the points of a curve are defined in terms of a parameter θ by relations

$$x = f_1(\theta), \quad y = f_2(\theta), \quad z = f_3(\theta), \quad t = f_4(\theta).$$

The line joining the points with parameters θ and $\theta + \delta\theta$ contains also the point

$$\left(\frac{f_1(\theta+\delta\theta)-f_1(\theta)}{\delta\theta}, \frac{f_2(\theta+\delta\theta)-f_2(\theta)}{\delta\theta}, \frac{f_3(\theta+\delta\theta)-f_3(\theta)}{\delta\theta}, \frac{f_4(\theta+\delta\theta)-f_4(\theta)}{\delta\theta}\right),$$

by the elementary definition of a straight line. Using a familiar limiting argument, we find that *the tangent at the point with parameter θ joins that point to the point* $(f_1'(\theta), f_2'(\theta), f_3'(\theta), f_4'(\theta))$, where dashes denote differentiations with respect to θ.

Theorem-examples.* 1. The osculating plane at a point P of the twisted cubic Γ contains the tangent at P.

2. If A, B, C, D are distinct points of Γ, the chords AD, BC are skew. [If they met, the plane containing them would meet Γ in four points.]

3. The tangents to Γ at two distinct points are skew.

4. Three osculating planes of Γ pass through a point $A(\xi, \eta, \zeta, \tau)$ not on the curve. If they osculate Γ at the points P, Q, R then the equation of the plane PQR is

$$x\tau - 3y\zeta + 3z\eta - t\xi = 0$$

and it passes through A.

5. If $XYZT$ is the tetrahedron of reference, then
 (i) XYZ is the osculating plane of Γ at X, and YZT is the osculating plane at T;
 (ii) XY is the tangent at X and ZT is the tangent at T;
 (iii) the points Y, Z do not lie on the curve.

* In these Theorem-examples, Γ is the curve $(\theta^3, \theta^2, \theta, 1)$.

6. If the curve is given, then a coordinate system in which its points have the parametric form $(\theta^3, \theta^2, \theta, 1)$ can be chosen as follows:

Let X, T be two *arbitrary* points of the curve; let the tangent at X meet the osculating plane at T in Y, and let the tangent at T meet the osculating plane at X in Z. Choose an arbitrary point of the curve as the unit point $(1, 1, 1, 1)$.

[Compare the preceding example.]

7. The chords of Γ which pass through the fixed point $(\alpha^3, \alpha^2, \alpha, 1)$ of the curve generate the quadric cone

$$\begin{vmatrix} x-y\alpha & y-z\alpha \\ y-z\alpha & z-t\alpha \end{vmatrix} = 0$$

or
$$(zx-y^2)-\alpha(xt-yz)+\alpha^2(yt-z^2) = 0.$$

[Eliminate β from the two equations given in § 3 (iii).]

8. *A unique chord of* Γ *passes through a fixed point not on the curve.* The chord through (ξ, η, ζ, τ) meets the curve in the two points whose parameters satisfy the equation

$$\begin{vmatrix} \theta^2 & \theta & 1 \\ \xi & \eta & \zeta \\ \eta & \zeta & \tau \end{vmatrix} = 0.$$

[Use § 3 (iii) and the fact that α, β are the roots of the equation

$$\theta^2-(\alpha+\beta)\theta+\alpha\beta = 0.]$$

9. The residual curve of intersection of two quadric cones with a common generator (on which the vertices therefore lie) is a twisted cubic.

[It meets an arbitrary plane in three points.]

10. A unique twisted cubic can be drawn through six points in general position.

[Consider cones projecting five of the points from the sixth.]

11. *The points of the twisted cubic curve* Γ *are projected from an arbitrary point* A *of* Γ *on to a plane* π *(not through* A*) in the points of a conic.*

[Take A as the point $(\alpha^3, \alpha^2, \alpha, 1)$ in Th.-ex. 7. The conic is the section of the cone obtained in that example by the plane π.]

12. The points of the tangents to Γ all lie on the surface whose equation is $(xt-yz)^2 = 4(zx-y^2)(yt-z^2)$.

13. The idea of an osculating plane can be extended to a more general curve by reasoning similar to that given for a tangent in the note. The equation of the osculating plane at the point with parameter θ is

$$\begin{vmatrix} x & y & z & t \\ f_1(\theta) & f_2(\theta) & f_3(\theta) & f_4(\theta) \\ f_1'(\theta) & f_2'(\theta) & f_3'(\theta) & f_4'(\theta) \\ f_1''(\theta) & f_2''(\theta) & f_3''(\theta) & f_4''(\theta) \end{vmatrix} = 0.$$

5. The quadrics through the curve. We have already seen that a twisted cubic lies on a number of quadrics; in fact, a quadric cone containing the curve can be drawn with its vertex at any point of the curve. We now find an expression for a general quadric drawn to contain the curve.

The equation of any quadric whatsoever is

$$ax^2+by^2+cz^2+dt^2+2fyz+2gzx+2hxy+2uxt+2vyt+2wzt = 0.$$

It contains the point $(\theta^3, \theta^2, \theta, 1)$ of the curve if

$$a\theta^6+b\theta^4+c\theta^2+d+2f\theta^3+2g\theta^4+2h\theta^5+2u\theta^3+2v\theta^2+2w\theta = 0.$$

This is a sextic equation, so that *an arbitrary quadric meets the curve in six points*. If the quadric contains the curve entirely, then this equation is an identity true for all values of θ. Hence

$$a = 0,\ 2h = 0,\ b+2g = 0,\ 2f+2u = 0,\ c+2v = 0,\ 2w = 0,\ d = 0.$$

Writing $b = -2g = A,\quad 2u = -2f = B,\quad c = -2v = C,$

we obtain the equation of a general quadric through the curve in the form

$$A(y^2-zx)+B(xt-yz)+C(z^2-yt) = 0.$$

This equation contains the three coefficients A, B, C and so *the equation of any quadric through the curve is linearly dependent on the equations of the three quadrics*

$$y^2-zx = 0,\quad xt-yz = 0,\quad z^2-yt = 0.$$

6. Involutions on the twisted cubic. As on the conic, so on the twisted cubic we may define correspondences between pairs of points. In particular, suppose that there is an involution on the cubic in which $P(\theta^3, \theta^2, \theta, 1)$, $Q(\phi^3, \phi^2, \phi, 1)$ are corresponding pairs. Then there is an equation of the form

$$a\theta\phi+b(\theta+\phi)+c = 0.$$

If (x, y, z, t) is an arbitrary point of the chord PQ, then

$$x-y(\theta+\phi)+z\theta\phi = 0,$$

$$y-z(\theta+\phi)+t\theta\phi = 0.$$

Eliminating $1:-(\theta+\phi):\theta\phi$ between these three equations, we find that *the chords joining the pairs of points in the given involution on the cubic form a regulus, generating the quadric whose equation is*

$$\begin{vmatrix} x & y & z \\ y & z & t \\ c & -b & a \end{vmatrix} = 0,$$

or

$$a(y^2-zx)-b(xt-yz)+c(z^2-yt) = 0.$$

Theorem-examples. 1. If a twisted cubic Γ lies on a quadric S, then each generator of one system meets Γ in a single point and each generator of the other system is a chord of Γ, cutting Γ in the pairs of points of an involution on the curve.

[The plane containing two generators meets Γ in three points, two on one generator and one on the other.]

2. A unique quadric can be drawn to contain a given twisted cubic Γ and a straight line which cuts Γ in one point.

[A unique quadric is defined by the given point, six other points of the cubic and two other points of the line.]

7. The line-coordinates of chords and tangents. The line-coordinates of the line joining the points $(\alpha^3, \alpha^2, \alpha, 1)$, $(\beta^3, \beta^2, \beta, 1)$ are

$$l = \alpha^3-\beta^3, \qquad m = \alpha^2-\beta^2, \qquad n = \alpha-\beta,$$

$$l' = \alpha^2\beta-\alpha\beta^2, \quad m' = \alpha\beta^3-\alpha^3\beta, \quad n' = \alpha^3\beta^2-\alpha^2\beta^3.$$

It is, however, more convenient to divide throughout by the factor $\alpha-\beta$ and to use the simpler form

$$l = \alpha^2+\alpha\beta+\beta^2, \quad m = \alpha+\beta, \qquad n = 1,$$

$$l' = \alpha\beta, \qquad m' = -\alpha\beta(\alpha+\beta), \quad n' = \alpha^2\beta^2.$$

The chords do not belong to any linear complex.

By putting $\beta = \alpha$ we obtain the line-coordinates of the tangent at α:

$$l = 3\alpha^2, \quad m = 2\alpha, \qquad n = 1,$$

$$l' = \alpha^2, \quad m' = -2\alpha^3, \quad n' = \alpha^4.$$

The tangents belong to the linear complex

$$l = 3l',$$

and also to the tetrahedral complex

$$mm' + 4nn' = 0.$$

Theorem-example. Four tangents of a twisted cubic belong to a general linear complex; in particular, four tangents of a twisted cubic meet an arbitrary line.

8. The cubic developable. The concept dual to a curve, regarded as a system of points, is a system of planes called a *developable*. In particular, the planes whose plane-coordinates can be expressed as cubic polynomials in a parameter θ are said to belong to a *cubic developable*. Thus the osculating planes

$$x - 3\theta y + 3\theta^2 z - \theta^3 t = 0$$

of the twisted cubic curve $(\theta^3, \theta^2, \theta, 1)$ define such a cubic developable. Clearly *three planes of a cubic developable pass through an arbitrary point of space.*

We do not enter into a detailed account of the properties of developables; the reader may obtain them, when required, as duals of the properties of curves.

EXAMPLES V

1. Find the osculating planes of the twisted cubic $(\theta^3, \theta^2, \theta, 1)$ which pass through the point $(18, 11, 6, 3)$.

2. Find the line-coordinates of the tangent to the twisted cubic $(\theta^3, \theta^2, \theta, 1)$ at the point $(1, 1, 1, 1)$, and verify that it meets the line whose coordinates are $(1, 4, 1, 1, 1, 1)$.

3. Obtain the coordinates of the twisted cubic which is the intersection (other than the line $x = y = 0$) of the quadrics

$$xt = yz, \quad x^2 + xz = y^2 - yt,$$

expressing your answer in the parametric form

$$x = \theta(1+\theta^2), \quad y = 1+\theta^2, \quad z = \theta(1-\theta^2), \quad t = 1-\theta^2.$$

Prove also that the osculating plane to this curve at the point

$$(10, 5, -6, -3)$$

is $13x - 14y + 11z - 2t = 0$.

4. Prove (do not merely verify) that the equation of the quadric which contains the curve $(\theta^3, \theta^2, \theta, 1)$ and the line joining the points $(1, 1, 1, 1)$, $(1, 0, 0, 1)$ is $y^2 - z^2 - zx + yt = 0$.

5. Prove that the equation of the cone whose vertex is the point $(1, 1, 1, 1)$ and which contains the cubic $(\theta^3, \theta^2, \theta, 1)$ is

$$y^2 + z^2 - yz - zx + xt - yt = 0.$$

6. Prove that the tangents of the twisted cubic $(\theta^3, \theta^2, \theta, 1)$ which belong to the linear complex

$$35l' + 5m' + n' - 25m + 24n = 0$$

touch the curve at the points whose parameters are 1, 2, 3, 4.

MISCELLANEOUS EXAMPLES V

1. Find equations for the chord joining the two points θ, ϕ of the twisted cubic curve $(\theta^3, \theta^2, \theta, 1)$.

Prove that the joins of pairs of points upon the curve which form the involution given by

$$a\theta\phi + b(\theta + \phi) + c = 0$$

are one system of generators of a quadric, and find the equation of the quadric.

L, M are any two points of the curve, and l, m are two lines passing respectively through L, M, but not meeting the curve again. Prove that l, m are met by one chord of the curve not passing through L and M. [P.]

2. Find expressions for the six coordinates of a chord of the twisted cubic $(\theta^3, \theta^2, \theta, 1)$, and obtain the equation of the surface generated by the chords of the curve which meet the line whose coordinates are $(\lambda, \mu, \nu, \lambda', \mu', \nu')$ in the form

$$\nu(zx - y^2)^2 + \lambda'(xt - yz)^2 + \nu'(yt - z^2)^2 + \mu'(xt - yz)(yt - z^2)$$
$$+ (\lambda - \lambda')(yt - z^2)(zx - y^2) - \mu(xt - yz)(zx - y^2) = 0. \quad \text{[P.]}$$

3. If P is any point in space, prove that there is a unique point Q which is the conjugate of P with respect to all quadrics containing a given twisted cubic.

Show also that, if P is a variable point on a fixed straight line, the locus of Q is another twisted cubic. [P.]

4. Prove that four tangents of a twisted cubic Γ meet a given line l.

If l' is the second transversal of these four tangents, show that the transversal to l, l' drawn from a general point P of Γ lies in the osculating plane at P. [P.]

5. Show that the equation of any quadric which passes through the curve $(\theta^3, \theta^2, \theta, 1)$ can be written in the form

$$A(zx - y^2) + B(xt - yz) + C(yt - z^2) = 0,$$

where A, B, C are constants.

Prove that this quadric contains two tangents of the curve; the parameters of the points of contact being the roots of the equation

$$A\theta^2+2B\theta+C = 0,$$

and hence or otherwise show that the equation of the cone projecting the curve from the point $\theta = \alpha$ is

$$(zx-y^2)-\alpha(xt-yz)+\alpha^2(yt-z^2) = 0. \qquad \text{[P.]}$$

6. A, B, C are three points on a twisted cubic; a generator through A of the quadric which contains the curve and the tangents to it at B, C meets the curve again at A'. The points B', C' are similarly determined. Prove that AA', BB', CC' are generators of a quadric containing the curve. [P.]

7. Prove that the lines joining pairs of corresponding points in an involution on a twisted cubic are generators of one system on a quadric Q, and that the intersection of the osculating planes at pairs of corresponding points of the involution are generators of one system on a quadric Q'.

Show that Q and Q' have four common generators, and identify them. [P.]

8. Prove that in general three quadrics with a generator in common meet in four further points. [From M.T. II.]

9. Prove the following properties of Γ the residual curve of intersection of two quadrics with a common generator:

(i) One chord of Γ passes through a general point of space.

(ii) If l is a line which meets Γ in one point P, the chords of Γ not passing through P which meet l generate a quadric surface.

(iii) If Γ is projected from an arbitrary point A of itself into a conic Γ' in a plane α, then those of its chords which meet an arbitrary line m that does not meet Γ project into the tangents to a conic Σ in α. Prove further that Σ is triangularly inscribed in Γ' (i.e. there exist triangles inscribed in Γ' and circumscribing Σ). [M.T. II.]

10. Prove that the locus of the poles, with respect to a linear complex, of the osculating planes of a twisted cubic Γ is another twisted cubic Γ', and that a tangent of Γ which belongs to the linear complex is also a tangent of Γ'.

Hence, by considering the linear complexes which contain four skew lines, show that if the lines are tangent to one cubic curve they are tangent to ∞^1 [an infinite number of] cubic curves. [M.T. II.]

11. Seven points are taken in space and a cubic curve is drawn through six of them. Show that the coordinates of the six points may be expressed in the form $(\theta^3, \theta^2, \theta, 1)$ for six values a, b, c, d, e, f of θ and the seventh point as $(p, 0, 0, 1)$. Prove that every quadric through the seven points passes through the further point $(q, 0, 0, 1)$, where $pq = abcdef$. [M.T. II.]

12. Prove that, if two quadrics have a generator l in common, their residual intersection is, in general, a twisted cubic curve which has l as a chord. Show also that any generator of either quadric belonging to the same system as l is a chord of the cubic curve.

Given two skew lines l, m, and four points A, B, C, D, prove that there is no twisted cubic through A, B, C, D having l and m as chords unless

$$l(A,B,C,D) = m(A,B,C,D),$$

and show that if this condition is satisfied there exists an infinity of cubics satisfying the requirements. [M.T. II.]

13. If the parameters of the points of contact of the osculating planes of the curve $(\theta^3, \theta^2, \theta, 1)$ which pass through an arbitrary point P are the roots of the equation

$$a_0\theta^3 + 3a_1\theta^2 + 3a_2\theta + a_3 = 0,$$

prove that the chord of the curve which passes through P joins the points whose parameters are given by

$$\begin{vmatrix} a_0\theta+a_1 & a_1\theta+a_2 \\ a_1\theta+a_2 & a_2\theta+a_3 \end{vmatrix} = 0. \qquad \text{[M.T. II.]}$$

14. Prove that any general line p which lies in two osculating planes of the twisted cubic $\Gamma(\theta^3, \theta^2, \theta, 1)$ meets the planes of reference XYZ, YZT in points whose coordinates may be written in the form $(3a, b, c, 0)$, $(0, a, b, 3c)$.

If p varies in such a way that its point of intersection with any one osculating plane of Γ describes a line, prove that its point of intersection with any other osculating plane likewise describes a line, and that the parametric representation may be so chosen that this variation of p is characterised by the condition $b = 0$.

Show also that p describes a quadric surface which touches every osculating plane of Γ and has three-point contact with Γ at each of two points. [M.T. II.]

15. The osculating planes at two points X, T of a twisted cubic curve Γ meet in a line l. A variable plane through l meets Γ in the points A, B, C. The lines BC, CA, AB meet l in P, Q, R respectively, and BQ meets CR in U. Prove that the locus of U is a twisted cubic curve Γ'.

Prove also that the osculating plane at U to Γ' meets Γ in A, in the harmonic conjugate of X with respect to (A', T) and in the harmonic conjugate of T with respect to (A', X), where A' is the harmonic conjugate of A with respect to (X, T). [M.T. II.]

16. The tangents are drawn to a twisted cubic at four points X, T, P, Q. Prove that, if the cross-ratio $(X, T; P, Q)$ is equianharmonic on the curve, then the four tangents have just one common transversal. [M.T. II.]

17. Show that any five given lines in space belong to a linear complex L, and that, if the five lines all touch the twisted cubic C, then every tangent of C belongs to L.

Prove that L is uniquely determined by C and contains no chords of C other than tangents. [P.]

18. Show that for any involution of pairs of points P, P' on a twisted cubic curve the joins PP' are the generators of one system on a quadric S.

A general plane π is met at Q, Q' by the generators of the other system through P, P' respectively. Prove that the lines QQ' concur at a point R, and show that, when π turns about a general fixed line l, the locus of R is a transversal line of the two lines PP' which meet l.

Determine linear equations between the (tangential) coordinates of the planes π and the (point) coordinates of the points R when P, P' correspond in the involution $\theta+\theta' = 0$ on the twisted cubic $(\theta^3, \theta^2, \theta, 1)$. [L.]

19. Prove that in a general plane π there lies a unique line l through which pass two osculating planes of the twisted cubic $(\theta^3, \theta^2, \theta, 1)$. If l' is the chord joining the points of contact of these osculating planes, prove that the osculating plane at a general point P of the curve contains the transversal from P to l and l'. [L.]

20. If a, b are skew lines in given planes α, β and A, B are points on a, b respectively, not on the line $(\alpha\beta)$, prove that a general point P lies on just one member of the family of twisted cubic curves $\{\Gamma\}$ through A and B, having a, b for tangent lines and α, β for osculating planes.

Show also that a general line is a chord of one and only one member of the system $\{\Gamma\}$. [L.]

21. Γ is the twisted cubic $(\theta^3, \theta^2, \theta, 1)$ on the quadric $xt = yz$. Prove that every generator of one system of the quadric meets Γ in one, and only one, point, while every generator g of the other system meets Γ in two points P, P' which coincide in points U, V respectively for two particular positions of g.

If the osculating planes of Γ at P, P' meet in a line l, prove that as g varies l generates a second quadric which has three-point contact with Γ at U, V. [L.; advanced.]

22. Prove that four tangents to a twisted cubic curve meet a general line of space, and that all the tangents to the cubic belong to a linear complex Σ. Show that the points of contact of tangents to the cubic which meet a general line of Σ form an equianharmonic set on the curve. [L.; M.Sc.]

23. The tangents to a twisted cubic curve Γ at four fixed points A, B, C, D are a, b, c, d. Prove that the four transversals drawn from any point P of Γ to the pairs of lines (a, b), (c, d), (AC, BD), (AD, BC) lie in one plane π.

Show also that, when P describes the curve Γ, the plane π turns about a fixed chord, whose intersections with Γ are the double points of the involution on Γ determined by the pairs (A, B) and (C, D). [L.; M.A.]

24. Prove that the locus in space of three dimensions determined by the vanishing of the 2-rowed minors of the matrix

$$\begin{pmatrix} x_{11} & x_{12} & x_{13} \\ x_{21} & x_{22} & x_{23} \end{pmatrix}$$

whose elements are general linear functions of the coordinates, is a curve of the third order, and that, by suitable choice of the coordinates, it can be expressed in the parametric form $(\theta^3, \theta^2, \theta, 1)$.

Show that one such curve passes through six points in general position.

P_1, P_2, P_3, P_4, P_5 are five general points in space of three dimensions, π is a general plane, and O is a general point of π. The line P_iP_j meets π in $P_{ij}(=P_{ji})$. The conic through O, P_{12}, P_{13}, P_{14}, P_{15} is denoted by s_1, and s_2, s_3, s_4, s_5 are similarly defined. Prove that the conics s_1, s_2, s_3, s_4, s_5 have two points in common besides O. [P.]

25. Find the equation of the unique quadric ψ through the twisted cubic $k(\theta^3, \theta^2, \theta, 1)$ containing the tangents to k at the points $X(1,0,0,0)$, $T(0,0,0,1)$.

Prove (i) that the osculating planes to k at X, T are the tangent planes to ψ at these points; (ii) that if π is the osculating plane to k at a variable point P of k, the pole Q of π with respect to ψ describes another twisted cubic k' through X and T; and (iii) that the tangent plane to ψ at P is the osculating plane to k' at Q. [L.]

26. A twisted cubic k and two of its tangents a, b are given. A transversal p of a and b varies in such a way that the two remaining of the four tangents of k which meet it are coincident. If p does not meet k, prove that it lies on a certain fixed quadric through a and b. [L.]

CHAPTER VI

SYSTEMS OF QUADRICS

1. Preliminary remarks. The reader will recall that, in discussing systems of conics called 'pencils', he had to consider various different cases according as the conics met in four distinct points, touched, had three-point or four-point contact, or had double contact. The work for quadrics is naturally even more complicated, and to obtain a classification which ensures that all possible cases are considered would carry us beyond the scope of this book. Instead, we shall name the types which occur most often, and *we assume throughout that a system has no greater complexity than is stated explicitly.*

Thus, if we were discussing a problem about two quadrics, we should regard it (for our purposes) as misplaced ingenuity to take them as two cones with a common vertex.

We use the standard notation

$$S \equiv ax^2+by^2+cz^2+dt^2+2fyz+2gzx+2hxy+2uxt+2vyt+2wzt,$$

with dashes for quadrics S', S'', etc.

For further details, see the admirable account in Dr J. A. Todd's *Projective and Analytical Geometry*, Chapter VI.

DEFINITION. The system of quadrics

$$\lambda S+\lambda' S' = 0,$$

where λ, λ' are parameters, is called a *pencil*, and the system

$$\lambda S+\lambda' S'+\lambda'' S'' = 0,$$

where λ, λ', λ'' are parameters, is called a *net*.

If Σ, Σ' are the tangential equations of two quadrics, the system

$$\lambda\Sigma+\lambda'\Sigma' = 0$$

is called a *tangential pencil* or a *range*.

Theorem-example. The quadrics of a pencil cut an arbitrary line in pairs of points in involution.

If we solve, for $x:y:z:t$, the quadratic equations $S = 0$, $S' = 0$ of two given quadrics and the linear equation $\pi = 0$ of an arbitrary plane, we obtain four solutions, so that the curve common to S and S' meets an arbitrary plane in four points; such a curve is called a *quartic curve*. All the quadrics of a given pencil pass through the quartic curve determined by any two of them. In particular cases this quartic curve may be *degenerate*, consisting of two conics (possibly 'coincident'), a conic and two lines (of which the lines may 'coincide'), a twisted cubic and a line, or four lines (perhaps not all distinct). We shall consider only the cases which arise most frequently, referring the reader to a more advanced treatise for a fuller discussion.

In the same way, if we solve, for $x:y:z:t$, the quadratic equations $S = 0$, $S' = 0$, $S'' = 0$ of three given quadrics, we get *in general* EIGHT POINTS, not necessarily distinct, and they are common to all quadrics of the net determined by S, S', S''. It is again easy to see that there are many cases of exception, and that a full discussion will be difficult. For example, if S'' belongs to the pencil determined by S, S', the intersection of the three quadrics is a *curve*; or if S, S', S'' are all drawn through a given conic R, then the intersection of the three quadrics consists of the *curve* R, together with a number of *isolated points*. Here, again, we shall assume that the intersection is no more complex than is explicitly prescribed. We shall also assume that the three quadrics which define a net are not in fact all members of a pencil.

Our first business is to discuss the properties of *pencils* of quadrics.

2. Polar properties of a pencil of quadrics. We begin with some properties which hold for all pencils of quadrics, taking the equation which defines the pencil in the standard form

$$S+\lambda S' = 0.$$

We use the notation of Chapter II, § 4, p. 36.

(i) *The polar planes of a point $P_1(x_1, y_1, z_1, t_1)$ with respect to the quadrics of the pencil pass through the fixed line whose equations are*

$$S_1 = 0, \quad S_1' = 0.$$

The proof is immediate. The line may be called *the polar line of* P_1 *with respect to the pencil.*

(ii) *If* P_1 *moves on a fixed straight line, then the polar line of* P_1 *with respect to the pencil generates a regulus.*

Let $P_2(x_2, y_2, z_2, t_2)$, $P_3(x_3, y_3, z_3, t_3)$ be two given points which define the fixed line. Then we can take

$$\mathbf{P}_1 \equiv \mathbf{P}_2 + \mu \mathbf{P}_3,$$

so that $$S_1 \equiv S_2 + \mu S_3, \quad S_1' = S_2' + \mu S_3'.$$

The polar line therefore generates a regulus on the quadric whose equation (found by eliminating μ between the equations $S_1 = 0$ and $S_1' = 0$) is

$$S_2 S_3' - S_3 S_2' = 0.$$

(iii) *The polar line of the line* $P_2 P_3$ *with respect to a variable quadric of the pencil generates (as the quadric varies) the other regulus on the quadric* $S_2 S_3' - S_3 S_2' = 0.$

The polar planes of P_2, P_3 with respect to the general quadric of the pencil are respectively

$$S_2 + \lambda S_2' = 0, \quad S_3 + \lambda S_3' = 0.$$

As λ varies, the line of intersection of these two planes describes the other regulus on the quadric $S_2 S_3' - S_3 S_2' = 0$.

(iv) *Two points which are conjugate with respect to two quadrics of a pencil are also conjugate with respect to all the quadrics of the pencil.*

The equations $$S_{12} = 0, \quad S_{12}' = 0$$

imply the equation $$S_{12} + \lambda S_{12}' = 0.$$

3. The cones of a pencil and the standard form of equation. If

$$S = 0, \quad S' = 0$$

are the equations of two quadrics, then, as we have seen, the equation

$$S + \lambda S' = 0,$$

where λ is a parameter, determines the pencil defined by them. Each quadric of the pencil contains the curve common to S and S', and one quadric of the pencil passes through an arbitrary point of space.

We must now find the *cones* in the pencil. We adopt the standard notation for the quadrics S, S', and it follows, by Chapter II, § 16, Theorem-example 2, p. 47, that the quadric $S+\lambda S' = 0$ is a cone if

$$\begin{vmatrix} a+\lambda a' & h+\lambda h' & g+\lambda g' & u+\lambda u' \\ h+\lambda h' & b+\lambda b' & f+\lambda f' & v+\lambda v' \\ g+\lambda g' & f+\lambda f' & c+\lambda c' & w+\lambda w' \\ u+\lambda u' & v+\lambda v' & w+\lambda w' & d+\lambda d' \end{vmatrix} = 0.$$

This is a quartic equation in λ, called the *discriminant* of the pencil. There are, *in general*, four roots of the equation and therefore *a general pencil of quadrics contains four cones.*

In the general case, no three vertices of these cones can lie on a line l. For then the involution cut on l by the quadrics of the pencil would have three self-corresponding points, and be identity. [Compare M., p. 30]. Hence the pencil would consist of quadrics with a common generator, namely l.*

Further, the vertices of the four cones of the general pencil cannot be coplanar. For let us suppose that, on the contrary, the four vertices lie in a plane π. The quadrics of the pencil cut π in conics through four fixed points, and the cones cut π in line-pairs (since their vertices are in π) passing through them. But there are only three line-pairs through four points in a plane when no three are collinear, and so the hypothesis is unsound.

Now take the four vertices of the cones (assumed distinct) as the vertices X, Y, Z, T of the tetrahedron of reference, and suppose that the values λ_1, λ_2, λ_3, λ_4 of λ give rise to the cones with vertices X, Y, Z, T respectively. By applying the result of Chapter II, § 16, Theorem-example 1, to these vertices in succession we obtain the relations

$$\begin{aligned} a+\lambda_1 a' &= 0, & h+\lambda_1 h' &= 0, & g+\lambda_1 g' &= 0, & u+\lambda_1 u' &= 0, \\ h+\lambda_2 h' &= 0, & b+\lambda_2 b' &= 0, & f+\lambda_2 f' &= 0, & v+\lambda_2 v' &= 0, \\ g+\lambda_3 g' &= 0, & f+\lambda_3 f' &= 0, & c+\lambda_3 c' &= 0, & w+\lambda_3 w' &= 0, \\ u+\lambda_4 u' &= 0, & v+\lambda_4 v' &= 0, & w+\lambda_4 w' &= 0, & d+\lambda_4 d' &= 0. \end{aligned}$$

* By taking the three vertices as $X(1,0,0,0)$, $T(0,0,0,1)$, $U(1,0,0,1)$, it may be proved that the quadrics would all be *cones* with their vertices on l.

Confining ourselves to the general case, in which λ_1, λ_2, λ_3, λ_4 are unequal, we have

$$f = f' = g = g' = h = h' = u = u' = v = v' = w = w' = 0,$$

and the equations of the quadrics are therefore

$$S \equiv ax^2 + by^2 + cz^2 + dt^2 = 0,$$
$$S' \equiv a'x^2 + b'y^2 + c'z^2 + d't^2 = 0.$$

By suitable choice of the unit point, we can put the equations in *the standard form for two general quadrics,*

$$S \equiv ax^2 + by^2 + cz^2 + dt^2 = 0,$$
$$S' \equiv x^2 + y^2 + z^2 + t^2 = 0.$$

[The latter reduction assumes, of course, that the quadric S' is not singular, in particular, that it is not one of the cones of the pencil.]

The equation of an arbitrary quadric of the pencil is then

$$(a+\lambda)x^2 + (b+\lambda)y^2 + (c+\lambda)z^2 + (d+\lambda)t^2 = 0,$$

where the cones are given by the values $-a$, $-b$, $-c$, $-d$ of λ. *The assumption that the cones are distinct implies that a, b, c, d are unequal.*

Theorem-examples. 1. The vertices of the four cones of a general pencil form a tetrahedron self-polar with respect to each quadric of the pencil.

[An immediate result of the forms just given for S and S'.]

2. There is one and only one tetrahedron self-polar with respect to all the quadrics of a general pencil.

[Each vertex of such a tetrahedron is the vertex of a cone of the pencil.]

3. If the reciprocal of the quadric $ax^2 + by^2 + cz^2 + dt^2 = 0$ with respect to the quadric $a'x^2 + b'y^2 + c'z^2 + d't^2 = 0$ is the quadric

$$a''x^2 + b''y^2 + c''z^2 + d''t^2 = 0,$$

then
$$a'^2 = aa'', \quad b'^2 = bb'', \quad c'^2 = cc'', \quad d'^2 = dd''.$$

4. The polar line p_1 of $P_1(x_1, y_1, z_1, t_1)$ with respect to the pencil

$$(a+\lambda)x^2 + (b+\lambda)y^2 + (c+\lambda)z^2 + (d+\lambda)t^2 = 0$$

is
$$\begin{cases} ax_1x + by_1y + cz_1z + dt_1t = 0, \\ x_1x + y_1y + z_1z + t_1t = 0. \end{cases}$$

5. The line-coordinates of p_1 (Th.-ex. 4) are given by

$$l_1 = (b-c)\,y_1 z_1, \quad m_1 = (c-a)\,z_1 x_1, \quad n_1 = (a-b)\,x_1 y_1,$$
$$l_1' = (a-d)\,x_1 t_1, \quad m_1' = (b-d)\,y_1 t_1, \quad n_1' = (c-d)\,z_1 t_1.$$

6. The polar line of an arbitrary point, with respect to the pencil of Th.-ex. 4, belongs to the tetrahedral complex

$$\frac{ll'}{a-d}+\frac{mm'}{b-d}+\frac{nn'}{c-d}=0.$$

4. Other types of pencil. Having examined the general pencil, consisting of quadrics through a non-degenerate quartic curve and including four cones with distinct vertices, we should now consider what other cases arise and obtain their classification. As explained before, we propose to evade the full rigour of the analysis, but we give a succession of common cases.

In order to be systematic, we shall study the ways in which the curve of degree four can break up. These will correspond to the partitions of the number four, namely,

$$[4], \quad [3,1], \quad [2,2], \quad [2,1,1], \quad [1,1,1,1],$$

so that, for example, [2, 2] will give rise to two curves of degree 2 (conics), not necessarily distinct. The general case already discussed arises from the partition [4]. The cases which we propose to consider are

(i) two distinct conics,
(ii) two 'coincident' conics,
(iii) four common generators.

Note that, in case (i), the planes of the conics must be distinct, but the conics themselves meet, on the line of intersection of their planes, in two (perhaps 'coincident') points. Also, in case (iii), the four lines must form a 'skew quadrilateral', say $ABCD$ (the four points not being coplanar), in which AB, CD are generators of one system and AC, BD generators of the other system on each quadric. It is easy to verify the necessity for these conditions.

Finally, that the reader may see where the difficulties lie in giving a full treatment, we quote two examples of pencils of quadrics which might not occur to him immediately:

(i) The cones with four given concurrent lines as generators.

(ii) The pairs of planes corresponding in a given involution of the planes of a pencil (for which the quartic curve consists of the line of intersection of the planes 'counted four times').

A full discussion of the subject must include all such features.

5. Equality of coefficients in the standard form. Consider what happens to the pencil defined by the quadrics

$$S \equiv ax^2 + by^2 + cz^2 + dt^2 = 0,$$

$$S' \equiv x^2 + y^2 + z^2 + t^2 = 0$$

in the cases when a, b, c, d are not distinct.

(i) *Suppose that $a = d$, but that b, c are different from them and from each other.* All the quadrics of the pencil $S + \lambda S' = 0$ contain their common curve; in particular, when $\lambda = -a$, the common curve lies on the quadric

$$(b-a)\,y^2 + (c-a)\,z^2 = 0,$$

which consists of the two planes

$$y\,\sqrt{(b-a)} \pm iz\,\sqrt{(c-a)} = 0.$$

The quartic curve common to the quadrics therefore breaks up into two conics, one in each of the above planes, and the conics meet in two points on the line of intersection of the planes.

(ii) *Suppose that $a = b = c \neq d$.* The quadrics meet where

$$x^2 + y^2 + z^2 = 0, \quad t^2 = 0,$$

and so the quartic curve breaks up into the conic, 'counted twice', in which the plane $t = 0$ cuts the cone $x^2 + y^2 + z^2 = 0$.

(iii) *Suppose that $a = d$, $b = c$ but that a, b are unequal.* Then each of the quadrics contains the four lines in which the two planes

$$x \pm it = 0$$

meet the two planes $\qquad y \pm iz = 0.$

The quartic curve breaks up into four straight lines forming a skew quadrilateral and comprising two generators of each system on all quadrics of the pencil.

We proceed to a discussion of these special cases.

6. The pencil of quadrics defined by two conics meeting in two distinct points. Let S_1, S_2 be two conics which have two distinct points X, T in common. Suppose that the tangents at X, T to S_1 meet in Y and that the tangents at X, T to S_2 meet in Z. If we take $XYZT$ as the tetrahedron of reference, the equations of S_1 and S_2 assume respectively the forms

$$z = 0, \quad by^2 + 2uxt = 0,$$

$$y = 0, \quad c'z^2 + 2u'xt = 0.$$

We can simplify these equations by the substitutions

$$y' = iy\sqrt{(b/2u)}, \quad z' = iz\sqrt{(c'/2u')},$$

and then (dropping dashes) we obtain the equations

$$z = 0, \quad y^2 - xt = 0,$$

$$y = 0, \quad z^2 - xt = 0.$$

To find the equations of the quadrics which pass through these two conics, write the general equation of a quadric in the form

$$Ax^2 + By^2 + Cz^2 + Dt^2 + 2Fyz + 2Gzx + 2Hxy + 2Uxt + 2Vyt + 2Wzt = 0.$$

This quadric meets $z = 0$ where $y^2 - xt = 0$, so that

$$A = D = H = V = 0, \quad B = -2U;$$

and it meets $y = 0$ where $z^2 - xt = 0$, so that

$$A = D = G = W = 0, \quad C = -2U.$$

The equation can therefore be taken as

$$y^2 + z^2 + 2\lambda yz - xt = 0,$$

where λ is arbitrary. The quadrics through the two conics thus belong to the pencil of which one quadric is

$$y^2 + z^2 - xt = 0$$

and another the plane-pair $\qquad yz = 0.$

To find the cones in the pencil, we recall that the equation of a cone must be expressible as a quadratic equation in *three* linear forms. Now x and t are two such forms, and the third must be found by choosing λ so that $y^2+2\lambda yz+z^2$ is its square. Hence $\lambda = \pm 1$, so that *there are two cones in this pencil*, and their equations are

$$\Omega_1 \equiv (y+z)^2 - xt = 0,$$
$$\Omega_2 \equiv (y-z)^2 - xt = 0$$

respectively. The vertices of the cones are the points V_1, V_2 given by the equations

$$y \pm z = 0, \quad x = 0, \quad t = 0,$$

so that

$$V_1 \equiv (0, 1, -1, 0), \quad V_2 \equiv (0, 1, 1, 0).$$

Hence *the vertices lie on the line YZ and separate the points Y, Z harmonically.*

Theorem-examples. 1. Any point of the line XT and any point of the line YZ are conjugate with respect to any quadric of the pencil.

2. If $S = 0$ is a given quadric and $\pi_1 = 0, \pi_2 = 0$ two given planes, then the equation $S + \lambda\pi_1\pi_2 = 0$ determines, as λ varies, a pencil of quadrics through the two conics $(S\pi_1)$, $(S\pi_2)$.

3. The residual intersection of two quadrics with a given common conic is another conic.

4. The discriminant (§ 3) of the pencil $y^2+z^2+2\lambda yz - xt = 0$ has roots $1, -1, \infty, \infty$, and so the plane-pair consisting of the planes of the two conics appears twice to make up the four cones of the pencil.

[Compare also the form given in § 5 (i).]

7. The pencil of quadrics touching along a given conic. Consider next the system of quadrics which is the limiting case of the pencil described in § 6 when the two conics 'coincide'. Suppose, in fact, that S is a given quadric and π a given plane meeting it in a (non-degenerate) conic C. We wish to examine those quadrics (if any) which touch S at all points of C.

Let T be the pole of π with respect to S. In the plane π take any triangle XYZ self-polar with respect to C. The tetrahedron $XYZT$ is self-polar with respect to S, and, by taking it as the tetrahedron of reference, we may write the equation of S (assumed non-singular) in the form

$$S \equiv x^2+y^2+z^2+t^2 = 0.$$

Since π is the plane $t = 0$, the equations which determine the conic C are

$$x^2+y^2+z^2+t^2 = 0, \quad t = 0.$$

We observe first of all that the equation of *any* quadric through C is of the form

$$S' \equiv x^2+y^2+z^2+dt^2+2uxt+2vyt+2wzt = 0.$$

If this quadric has the same tangent plane as S at each point of the conic C, then π is the polar plane of T with respect to S' also. The two planes

$$t = 0, \quad ux+vy+wz+dt = 0$$

thus coincide, so that $\quad u = v = w = 0.$

The equation of S' can therefore be taken in the form

$$x^2+y^2+z^2+\lambda t^2 = 0,$$

where λ is a parameter. As λ varies, we obtain a pencil of quadrics, defined by S and the 'repeated plane' consisting of π counted twice.

[Compare the result (in converse form) given in § 5 (ii).]

Such quadrics are said to have *ring contact*.

Theorem-examples. 1. There is just one cone in a pencil of quadrics having ring-contact. The other three cones of the general case all reduce to the repeated plane of the conic of contact.

2. If $S = 0$ is an arbitrary quadric and $\pi = 0$ an arbitrary plane, the equation $S+\lambda\pi^2 = 0$ defines a pencil of quadrics with ring contact round the conic $(S\pi)$.

3. If $S_i \equiv x^2+y^2+z^2+\lambda_i t^2$, then the reciprocal of the quadric $S_1 = 0$ with respect to the quadric $S_2 = 0$ is the quadric $S_3 = 0$, where $\lambda_2^2 = \lambda_1\lambda_3$. Discuss what happens if λ_1 or λ_2 vanishes.

8. The pencil of quadrics with four common generators. Let a, b be two given skew lines and c, d two other skew lines, each of which meets both a and b. Consider those quadrics which have the four lines as generators.

Let $X \equiv (ac)$, $Y \equiv (ad)$, $Z \equiv (bc)$, $T \equiv (bd)$. With X, Y, Z, T as tetrahedron of reference, the given lines are

$$a: \quad z = t = 0; \qquad b: \quad x = y = 0;$$

$$c: \quad y = t = 0; \qquad d: \quad x = z = 0.$$

The equation of any quadric containing each of these four lines is of the form

$$xt+\lambda yz = 0,$$

so that the quadrics belong to a pencil.

Theorem-examples. 1. There are no (non-degenerate) cones in the pencil. The plane-pairs $xt = 0$ and $yz = 0$ each count twice to make up the four cones.

2. Two quadrics with *two* common skew generators meet residually in two generators of the opposite system on each.

3. If $S_i \equiv xt + \lambda_i yz$, then the reciprocal of the quadric $S_1 = 0$ with respect to the quadric $S_2 = 0$ is the quadric $S_3 = 0$, where $\lambda_2^2 = \lambda_1 \lambda_3$. Discuss what happens if λ_1 or λ_2 vanishes.

4. The equation for the quadrics of the pencil can be transformed to the form

$$\xi^2 + \tau^2 + \lambda(\eta^2 + \zeta^2) = 0.$$

[Compare the result of § 5 (iii).]

9. Tangential pencils. The properties of tangential pencils are, of course, dual to those of ordinary pencils, and we shall not give the results in detail. We state, however, one or two theorems for the general tangential pencil whose equation is

$$\Sigma + \lambda \Sigma' = 0,$$

where

$$\Sigma \equiv al^2 + bm^2 + cn^2 + dp^2,$$

$$\Sigma' \equiv l^2 + m^2 + n^2 + p^2.$$

(i) One quadric of the tangential pencil touches an arbitrary plane.

(ii) The tangential pencil contains four 'disc quadrics' (or, speaking loosely, conics) given by the values $-a, -b, -c, -d$ of λ.

(iii) The poles of an arbitrary given plane with respect to the quadrics of the tangential pencil all lie on a fixed line; if the given plane varies in a pencil, the line generates a regulus and the polar lines of the line with respect to the individual quadrics of the pencil generate the complementary regulus.

A fuller treatment is inserted in Chapter VII, § 11, p. 123, at the point where further details are required for metrical developments.

10. Nets of quadrics; associated points. We have defined a *net* of quadrics as a system given by the equation

$$\lambda S + \lambda' S' + \lambda'' S'' = 0,$$

where $\lambda, \lambda', \lambda''$ are parameters. One such system is determined by the quadrics through seven *arbitrary* points of space; for the equation of a quadric depends linearly on the ten arbitrary constants $a, b, c, d, f, g, h, u, v, w$, while the passing of a quadric through each of the seven points imposes, successively, seven linear conditions upon the constants. In the general case these conditions are independent and serve to reduce the number of disposable constants from ten to three.

In these circumstances we have the following important theorem:

All the quadrics through seven given points in arbitrary position in space pass through an eighth point determined by them.

For the equation $$\lambda S+\lambda' S'+\lambda'' S''=0$$

which determines the system shows that all these quadrics pass through the points at which S, S' and S'' all vanish, and there are eight such points.

Eight points so related are said to be *associated*.

It may be useful to remark that the quadrics S, S', S'' used to define a net need not be non-singular. In particular, three quadrics which meet in eight distinct points define an associated set, even though one or more of the quadrics is, say, a plane-pair.

The use of the theorem of eight associated points enables us to establish the property given in Chapter I, Illustration 3, p. 18:

ILLUSTRATION. *Given that $XYZT$, $ABCD$ are two tetrahedra with the property that A, B, C, D lie respectively in the planes YZT, ZXT, XYT, XYZ while X, Y, Z lie in the planes BCD, CAD, ABD. To prove that T lies in the plane ABC.*

The three quadrics defined by the plane-pairs

$$S\equiv YZTA,\ BCDX,$$
$$S'\equiv ZXTB,\ CADY,$$
$$S''\equiv XYTC,\ ABDZ$$

all contain the eight vertices of the tetrahedra, which therefore form an associated set. Hence any quadric through A, B, C, D, X, Y, Z contains T. Thus the plane-pair

$$XYZD,\quad ABC$$

contains T, and so T lies in one or other of the planes XYZ, ABC. But T cannot lie in the plane XYZ, otherwise $XYZT$ would not be a tetrahedron, and so T lies in the plane ABC.

Theorem-examples. 1. The eight points consisting of a given point, its four harmonic inverses with respect to the vertices and opposite faces of a given tetrahedron, and its three harmonic inverses with respect to the opposite edges of the tetrahedron form an associated set.

[If the point (ξ, η, ζ, τ) lies on the quadric $ax^2+by^2+cz^2+dt^2 = 0$, so do the points $(\pm\xi, \pm\eta, \pm\zeta, \pm\tau)$.]

2. The quadrics of the net

$$\lambda(x^2+y^2+z^2+t^2)+\mu yt+\nu zt = 0$$

have in common a conic and two points not on the conic.

[The conic is given by the equations $t = 0$, $x^2+y^2+z^2 = 0$.]

3. The quadrics of the net

$$\lambda(y^2-zx)+\mu(xt-yz)+\nu(z^2-yt) = 0$$

have in common a twisted cubic.

4. The quadrics of the net

$$\lambda yz+\mu xt+\nu zt = 0$$

have three lines in common.

These examples illustrate the variety of cases which can arise when we consider a net of quadrics. Even an introductory analysis would require a fuller treatment than we propose for this book.

MISCELLANEOUS EXAMPLES VI

1. If $S = 0$ is a quadric and $\alpha = 0$, $\beta = 0$ are two planes, what relation exists between the quadrics $S = 0$, $S = \alpha\beta$?

The faces of a tetrahedron $XYZT$ are the planes $x = 0, y = 0, z = 0, t = 0$. Obtain the equations of two quadrics S_1, S_2 of which S_1 touches the plane XYZ, XYT at Z and T and S_2 touches the planes ZTX, ZTY at X and Y.

Show that, if S_1 and S_2 touch one another at P and Q, the line PQ must intersect XY and ZT. [P.]

2. If $S = 0$, $S' = 0$ are the point equations of two quadrics, show that the locus of the poles of a fixed plane with regard to quadrics of the family $S+\lambda S' = 0$ is a twisted cubic and that the polar lines of a fixed line generate a quadric surface. [P.]

3. Two coplanar triangles ABC and DEF are in perspective in that order. If ABC and EFD are also in perspective, prove that ABC and FDE are in perspective.

$1, 2, \ldots, 6$ are six general points in space and (12), (13), ... are the lines joining them. Q is the quadric through (14), (25), (36), Q' the quadric through (15), (26), (34), and Q'' the quadric through (16), (24), (35). Prove that Q, Q', Q'' belong to a pencil. [M.T. II.]

4. Prove that through any two plane sections of a quadric two quadric cones can be drawn.

α, β, γ are the sections of a quadric by three planes meeting in O. Through β and γ, γ and α, α and β quadric cones are drawn with vertices A_1 and A_2, B_1 and B_2, C_1 and C_2 respectively. Show that these vertices all lie on the plane which is the polar of O with regard to the quadric, and that A_1 and A_2, B_1 and B_2, C_1 and C_2 are opposite vertices of a quadrilateral. State the dual theorem. [M.T. II.]

5. Prove that the polar planes of a point P with respect to the four cones of a given pencil of quadrics form a pencil whose cross-ratio is independent of the position of P. State the dual of this result.

Show that all the planes which touch two given conics in space touch, in general, each of two other fixed conics, and that the four points of contact of any one of the planes lie on a line and form a range of constant cross-ratio.

If one of the given conics is a point-pair, show that the planes touch only one further fixed conic and identify the plane of this conic. [M.T. II.]

6. Prove that, in general, the equations of any two quadrics can be reduced simultaneously to the forms

$$ax^2+by^2+cz^2+dt^2 = 0, \quad a'x^2+b'y^2+c'z^2+d't^2 = 0.$$

The planes α, β, γ, δ and the planes α', β', γ', δ' are the faces of two tetrahedra, each of which is self-polar with respect to a quadric S. Prove that there is another quadric S' with respect to which the tetrahedra with faces $\alpha', \beta, \gamma, \delta$ and with faces $\alpha, \beta', \gamma', \delta'$ are each self-polar, and that the common points of S and S' lie on two planes harmonically separating the planes α, α'. [M.T. II.]

7. Find the nature of the intersection of the quadrics whose equations are

$$xt-yz = 0, \quad ab(x^2-t^2)+(ax-bt)(z+aby)+(a^2-b^2)yz = 0.$$

[M.T. II.]

8. S and S' are two given quadrics. The polar planes of the point P with respect to S and S' intersect in the line p. Prove that p meets the faces of the self-polar tetrahedron of S and S' in four points whose cross-ratio is constant for all positions of P. [M.T. II.]

9. Show that the equation of any quadric of a pencil whose common curve is a skew-quadrilateral can be written in the form

$$axt+fyz = 0.$$

A quadric S_1 is such that each of its generators is the polar with regard to a quadric S_2 of a generator of the same system. Show that the common curve of the two quadrics is a skew quadrilateral and that their equations can be reduced to the form

$$xt+yz = 0, \quad xt-yz = 0.$$

[M.T. II.]

10. Prove that two conics S, T, in different planes α, β, which meet the line $(\alpha\beta)$ at the same two points C, D are in perspective from each of two points V_1, V_2.

If S, T are sections of a given quadric Φ, prove that V_2 is the pole with respect to Φ of the plane joining V_1 to the intersection of α with the polar plane of V_1 with respect to Φ; and show that, if S remains fixed while V_1 describes a line, then in general V_2 describes a conic. [L.]

11. Prove that the point equation of any general quadric S can be written in the form $XT = YZ$, where X, Y, Z, T are linear functions of the point coordinates; and state the relation of the lines $X = T = 0$ and $Y = Z = 0$ with respect to S.

Two skew lines l, m are polar lines with respect to a quadric of the pencil determined by two given quadrics S_1, S_2. Show that there are two quadrics each of which touches one of S_1, S_2 at its intersections with l and the other at its intersections with m. [L.]

12. Two quadrics S, S' are given by

$$y^2 - yz - zx - xt - yt = 0,$$

$$z^2 + t^2 - yz + xt - yt + 2zt = 0.$$

By discussing their discriminant $| S - \lambda S' | = 0$, prove that their intersection consists of a common generator and a twisted cubic. [L.; advanced.]

13. A twisted cubic curve Γ passes through the vertices A, B, C, D of a tetrahedron self-polar with respect to a quadric S. Prove that any general point of Γ is one vertex of a tetrahedron inscribed in Γ and self-polar with respect to S. [L.]

14. Γ is a given conic whose plane is π; l, m are given skew lines, neither of which lies in π. A variable plane α through l meets m in a point P and π in a line λ; Q is the pole of λ with respect to Γ. Prove that the line PQ generates one regulus of a quadric which touches the plane π. [From L.; advanced.]

15. Prove that a unique quadric surface can be drawn through a twisted cubic curve and any two points of space not on the same chord of the curve.

If a conic S has three points in common with a twisted cubic curve Γ, show that the chords of Γ which meet S generate a quadric surface. [From L.; M.A.]

16. Prove that those quadrics which pass through seven points of general position in space have also an eighth point in common. [The eight points are called the *base points* of the system of quadrics.]

Prove that the polar planes of a point P with respect to all the quadrics have in general a single common point P', and that the point common to the polar planes of P' is P.

When P describes any curve c, the point P' describes the curve c' *conjugate* to c. Show that, when c is a line, c' is in general a twisted cubic; but that, if c is a line joining two base points, the curve conjugate to c is c itself.

Show also, conversely, that if a line is a self-conjugate curve it must join two of the base points. [M.T. II.]

17. Prove that the quadric surfaces for which a given tetrahedron $ABCD$ is self-polar determine an involution on any line which either passes through a vertex or meets a pair of opposite edges of the tetrahedron $ABCD$.

Two tetrahedra $ABCD$ and $ABEF$ have the edges CD and EF in the same line. Show that all quadrics of the system for which $ABCD$ and $ABEF$ are self-polar tetrahedra meet the line CD at the same two points M, N, and touch fixed planes at M and N; and that the quadrics of this system which pass through any given point P of AB form a pencil with four common generators.

If a line meet AB and CD at T, U respectively, and a conic Σ for which the triangle UAB is self-polar meet TU at R and S, prove that there is a conic Σ' which touches TM, TN at M, N respectively, and also passes through R and S; and show that there is just one quadric through Σ and Σ' for which the tetrahedra $ABCD$ and $ABEF$ are self-polar. [L.; M.A.]

18. Two lines l, m are given, and also a quadric ψ for which each of l, m is the polar line of the other. Variable planes are drawn through l, m respectively, to meet ψ in conics s, t. Show that the vertices of the cones through s, t lie on a fixed quadric ϕ which meets ψ in four straight lines, and show also that the relation between ϕ and ψ is symmetrical. [L.]

CHAPTER VII

APPLICATIONS TO EUCLIDEAN GEOMETRY

THE results established in the earlier chapters can be interpreted metrically in ways closely analogous to those used in plane geometry. The space with which we have been dealing, defined by four complex homogeneous coordinates, is called *projective space*; the ordinary space of metrical geometry (in the commonly accepted sense of the term) is called *Euclidean space*.

As in plane geometry, we need a justification for the use of 'infinite' and 'imaginary' elements. The extension from two to three dimensions is, however, so immediate that we shall take it for granted in what follows.

[See, for example, M., Chapters XI and XII.]

We assume that the reader has had an introductory course of solid geometry involving the use of three real Cartesian coordinates, usually denoted by x, y, z. Our needs here are slight: a knowledge of the equations for the plane, straight line and sphere, together with the conditions of parallelism and perpendicularity, and a few metrical definitions for quadrics will probably be enough—though, as in all such cases, a *sound* knowledge will help towards a deeper understanding of the interpretations.

1. Homogeneous Cartesian coordinates; the plane at infinity. We begin by expressing the coordinates x, y, z of ordinary Cartesian solid geometry in the homogeneous form x/t, y/t, z/t. This in itself adds nothing new, but it is suggestive in two ways. First, we can make the equations of Cartesian geometry homogeneous, thereby establishing a similarity with the equations of projective space; and, secondly, we are led to consider what happens if t is allowed to assume the value zero.

If, in fact, x, y, z are non-zero while t itself vanishes, then the homogeneous Cartesian coordinates x/t, y/t, z/t have 'infinite' values. There are, of course, no points in Euclidean space with such coordinates; but we suppose that the space is augmented, as it were, by a system of points, which we regard as lying in a plane, the *plane at infinity*, given by the equation $t = 0$. The augmentation by these points and by the points with complex coordinates gives an 'augmented Euclidean space' which, if we abandon metrical ideas, becomes the complex projective space.

Suppose, then, that we have a figure in projective space whose properties are reflected in terms of homogeneous coordinates x, y, z, t. We can obtain a metrical interpretation (by rules to be determined) in which the real numbers x/t, y/t, z/t are taken as Cartesian coordinates, the plane $t = 0$ being the 'plane at infinity'. By suitable choice of the system of reference in the projective space, an *arbitrary* plane can be selected and interpreted as the 'plane at infinity', so that one projective figure is capable of many metrical interpretations.

2. Parallel straight lines. The reader will be familiar with the formula

$$\frac{x-x_1}{l} = \frac{y-y_1}{m} = \frac{z-z_1}{n}$$

of ordinary Cartesian solid geometry, in which the coordinates x, y, z of a variable point of a given line are expressed in terms of the coordinates x_1, y_1, z_1 of one of its points and three numbers l, m, n called the *direction-cosines* of the line. In particular, the ratios $l:m:n$ determine the *direction* of the line, and all parallel straight lines are characterised by equal direction-cosines.

In terms of homogeneous Cartesian coordinates, the equations of the straight line are

$$\frac{(x/t)-(x_1/t_1)}{l} = \frac{(y/t)-(y_1/t_1)}{m} = \frac{(z/t)-(z_1/t_1)}{n}$$

or

$$\frac{t_1x-x_1t}{l} = \frac{t_1y-y_1t}{m} = \frac{t_1z-z_1t}{n},$$

and these equations are satisfied by the coordinates

$$(x_1+\lambda l,\ y_1+\lambda m,\ z_1+\lambda n,\ t_1)$$

for all values of λ. The straight line therefore passes through the point (x_1, y_1, z_1, t_1)—the case, $\lambda = 0$—and also through the point $(l, m, n, 0)$ in the 'plane at infinity'—the case, $\lambda = \infty$.

But the ratios $l:m:n$ are the same for all parallel lines, and so we obtain the rule that *straight lines which meet in the 'plane at infinity' are to be interpreted as parallel.*

Theorem-examples. 1. A plane can be drawn to contain each of two given parallel straight lines.

2. A unique line can be drawn through a given point parallel to a given line.

[The line joins the given point to the 'point at infinity' on the given line.]

3. Parallel planes. Any two planes in projective space necessarily meet in a line. Consider the case when that line lies in the plane to be chosen as the 'plane at infinity'. An arbitrary plane then meets the given planes in two lines which, meeting in the 'plane at infinity', are therefore parallel. Hence two planes which meet in the 'plane at infinity' are met by an arbitrary plane in two parallel lines. But this is a characteristic property of two parallel planes, familiar in ordinary Euclidean geometry. We therefore have the result that *two planes whose line of intersection lies in the plane to be chosen as the 'plane at infinity' are to be interpreted as parallel.*

4. Length or distance. The introduction of length for figures in space is governed by ideas exactly analogous to those for a plane. We have the following rule (compare M., Chapter XI):

Suppose that P, U, A are three points lying on a line which meets the 'plane at infinity' in K. Then *in interpreting results from projective space to Euclidean space, we interpret the cross-ratio*

$$(P, U, A, K)$$

(*where K is in the plane to be chosen as the 'plane at infinity'*) *as the ratio of the lengths AP/AU.*

In particular, if $(P, U, A, K) = -1$, so that the four points form a harmonic range, then A is the middle point of UP; in other words, P is the *reflection* of U with respect to the point A.

ILLUSTRATION 1. *The centroid of a tetrahedron.* Let $XYZT$ be a given tetrahedron and π a given plane. Denote by L, M, N, L', M', N' the harmonic conjugates, with respect to the relevant vertices of the tetrahedron, of the points in which the edges XT, YT, ZT,

YZ, ZX, XY respectively meet the plane π. We prove that *the lines* LL', MM', NN' *are concurrent.*

Suppose that the equation of the plane π, referred to the tetrahedron $XYZT$, is

$$lx + my + nz + pt = 0.$$

It is easy to obtain the coordinates

$$L(p, 0, 0, l), \quad M(0, p, 0, m), \quad N(0, 0, p, n),$$
$$L'(0, n, m, 0), \quad M'(n, 0, l, 0), \quad N'(m, l, 0, 0),$$

and to see that the point

$$G(l^{-1}, m^{-1}, n^{-1}, p^{-1})$$

lies on each of the lines LL', MM', NN'.

Interpretation in Euclidean space. Suppose that π is taken as the 'plane at infinity'. We obtain at once the interpretation that *the lines joining the middle points of the opposite edges of a tetrahedron are concurrent.* This point is called the *centroid* of the tetrahedron.

Further, suppose that XG meets the plane YZT in the point X_1 and the plane π in K_1. In terms of the symbols of X and G, we have

$$\mathbf{X}_1 \equiv -l^{-1}\mathbf{X} + \mathbf{G},$$
$$\mathbf{K}_1 \equiv -4l^{-1}\mathbf{X} + \mathbf{G}.$$

Hence

$$(X, G, X_1, K_1) = (\infty, 0, -l^{-1}, -4l^{-1})$$
$$= 4,$$

so that, in the interpretation in Euclidean space,

$$X_1X/X_1G = 4.$$

Theorem-example. X_1 is the centroid of the triangle YZT.

5. First properties of quadrics. A quadric meets the plane π to be chosen as the 'plane at infinity' in a conic. We call the pole of π the *centre* of the quadric; any chord through the centre is a *diameter* and each diameter is bisected at the centre. The tangent cone from the centre is called the *asymptotic cone* of the quadric; a quadric and its asymptotic cone meet the 'plane at infinity' in the same conic.

It may happen, however, that the quadric touches the 'plane at infinity', so that the conic of intersection is degenerate. In that case the pole of the 'plane at infinity' is itself 'at infinity'; we say that the quadric is *non-central*. Such quadrics are called *paraboloids*. The central quadrics are called *ellipsoids* and *hyperboloids*, the difference between these two classes depending on considerations of reality into which we do not enter here.

6. The sphere. Taking the hint from the treatment of a circle in a plane, consider the sphere whose equation in homogeneous Cartesian coordinates is

$$(x/t)^2+(y/t)^2+(z/t)^2+2u(x/t)+2v(y/t)+2w(z/t)+d = 0$$

or

$$x^2+y^2+z^2+2uxt+2vyt+2wzt+dt^2 = 0.$$

It meets the plane $t = 0$ in the conic given by the equations

$$x^2+y^2+z^2 = 0, \quad t = 0,$$

and *these equations do not contain the constants* u, v, w, d *which particularise the sphere.* We are therefore led to extend the *circular points* of plane geometry to an *absolute conic* in space, according to the following rule:

In an interpretation of projective space in which a given non-degenerate conic Ω *is to be chosen as the absolute conic (its plane necessarily being chosen as the 'plane at infinity'), any quadric passing through* Ω *is to be interpreted as a sphere.*

An arbitrary plane of space meets the conic Ω in two points which are the circular points for the geometry of that plane. In particular, a plane section of a 'sphere' is a conic through the circular points in its plane, i.e. a 'circle'.

It is possible to deduce from this basis, with certain obvious results obtained by analogy with plane geometry, the leading properties of spheres; in particular, we could prove the 'equal radius' property which is the usual definition. The author, however, feels that there is no great point in an elaborate development of the theory, and attention is confined here to results which seem to arise naturally in the spirit of the book as a whole.

In the same way, we shall not enter into a discussion on angles in general, but we content ourselves with the case of the right angle, to which we now proceed.

7. The right angle. We begin, as usual, with Euclidean geometry expressed by means of homogeneous Cartesian coordinates. The equations of a line whose direction is given by direction-cosines l, m, n are

$$\frac{(x/t)-\alpha}{l} = \frac{(y/t)-\beta}{m} = \frac{(z/t)-\gamma}{n}$$

or

$$\frac{x-\alpha t}{l} = \frac{y-\beta t}{m} = \frac{z-\gamma t}{n},$$

and this line meets the plane $t = 0$ in the point $(l, m, n, 0)$. This point is independent of α, β, γ, as we should expect from the fact that parallel lines all cut the 'plane at infinity' in the same point.

It is well known that the condition for two directions

$$(l_1, m_1, n_1), \quad (l_2, m_2, n_2)$$

to be perpendicular is

$$l_1 l_2 + m_1 m_2 + n_1 n_2 = 0,$$

and this is precisely the condition for the two points $(l_1, m_1, n_1, 0)$, $(l_2, m_2, n_2, 0)$ to be conjugate with respect to the absolute conic

$$x^2+y^2+z^2 = 0, \quad t = 0.$$

Hence *if two points are conjugate with respect to the conic to be chosen as the absolute conic, then two lines, one through each of the points, are to be interpreted as perpendicular.*

An immediate extension leads to the *normal* to a plane. Suppose that a given plane λ meets the 'plane at infinity' in a line l, and that L is the pole of l with respect to the absolute conic. Then L and any point of l are conjugate with respect to the absolute conic; and so we have the interpretation that *every line through L is perpendicular to every line in the plane.* The point L therefore defines the direction of lines perpendicular or *normal* to the plane. For example, a unique line can be drawn through a given point perpendicular to a given plane. Again, two planes are perpendicular if their normals are perpendicular, and so *two planes which meet the 'plane at infinity' in lines conjugate with respect to the absolute conic are to be interpreted as perpendicular.*

ILLUSTRATION 2. *The orthogonal tetrahedron. $XYZT$ is a tetrahedron in which YT is perpendicular to ZX and ZT is perpendicular to XY. To prove that XT is perpendicular to YZ.* The four faces of the tetrahedron meet the 'plane at infinity' in the sides of a quadrilateral of which two pairs of opposite vertices are conjugate with respect to the absolute conic. It is a standard theorem of plane geometry that the remaining vertices are also conjugate, and the result is immediate.

Such a tetrahedron is said to be *orthogonal.*

ILLUSTRATION 3. *The configuration of the cube.* We begin with the system of three desmic tetrahedra $XYZT$, $ABCD$, $PQRS$ described in Chapter I, Illustration 5 (p. 20). It will be remembered that any two of the tetrahedra are in fourfold perspective from the vertices of the third. In particular, to take the case in which we are interested, AS, BR, CQ, DP meet in X; AR, BS, CP, DQ meet in Y; AQ, BP, CS, DR meet in Z; and AP, BQ, CR, DS meet in T.

Note first that, with the coordinate system used before, each of the tetrahedra is self-conjugate with respect to the quadric

$$x^2+y^2+z^2+t^2 = 0.$$

This is clear when we remember that $XYZT$ is the tetrahedron of reference and that the vertices of the two other tetrahedra are $(1, -1, -1, 1)$, $(-1, 1, -1, 1)$, $(-1, -1, 1, 1)$, $(1, 1, 1, 1)$ and $(-1, 1, 1, 1)$, $(1, -1, 1, 1)$, $(1, 1, -1, 1)$, $(1, 1, 1, -1)$ respectively.

Consider, then, the section of the configuration by the plane XYZ, which cuts the quadric just determined in a conic Ω. In the first place, the triangle XYZ is self-conjugate with respect to Ω. Moreover, the pairs of opposite edges of the two tetrahedra $ABCD$, $PQRS$ cut the plane XYZ in pairs of points conjugate with respect to Ω.

Let us seek an interpretation in which Ω is taken as the absolute conic, so that XYZ is the 'plane at infinity'. In virtue of the perspectives from X, Y, Z, the lines

$$AS, BR, CQ, DP; \quad AR, BS, CP, DQ; \quad AQ, BP, CS, DR$$

are grouped in three parallel sets, any line of one set being perpendicular to any line of each of the other sets. The lines AP, BQ, CR, DS are therefore the diagonals of a rectangular box. Also the pairs of lines BC, AD; CA, BD; AB, CD; QR, PS; RP, QS; PQ, RS (the first set of three would be sufficient) are perpendicular, so that each face of the box is a rectangle with perpendicular diagonals, i.e. a square. The two desmic tetrahedra $ABCD$, $PQRS$ therefore define the vertices of a *cube*, the other tetrahedron of the set being given by the centre of the cube and the 'points at infinity' in the directions of the sides.

Note. It may be helpful to indicate a sequence of increasing specialisation for the figure $ABCDPQRS$. If the absolute conic is in general position in the plane XYZ, the figure is a parallelepiped; if the conic is chosen to have XYZ as a self-conjugate triangle, the parallelepiped is rectangular; if the particular conic described above is chosen, the figure is a cube.

8. Further properties of spheres. Let S be a given quadric, met by a plane π in a non-degenerate conic Ω, and let O be the pole of π. If P is an arbitrary point of S, then the tangent plane at P meets π in a line l which is the polar line of OP with respect to S. The line OP thus meets π in the pole of l with respect to the conic Ω.

In an interpretation in which Ω is the absolute conic, S is a sphere and O its centre. The result just proved shows that the radius OP is perpendicular to the tangent plane at P.

We suggest in the form of Theorem-examples some properties of two spheres which are suitable for treatment by the present methods.

Theorem-examples. 1. Two spheres intersect in a circle.

[Two quadrics which meet in a given conic meet residually in a further conic. Of course, the circle of intersection may be 'imaginary'.]

2. The line joining the centres of two intersecting spheres passes through the centre of their common circle and is perpendicular to the plane of the circle.

3. If there is one point common to two spheres such that the radii to it are perpendicular, then the same is true of all common points.

DEFINITIONS. The plane which contains the circle (real or 'imaginary') common to two spheres is called their *radical* plane. This plane has the property, not to be proved here but deducible from the case of two dimensions, that the tangents from any point of it to the two spheres are equal in length.

The spheres through a given circle (real or 'imaginary') are said to form a *coaxal system*. They form a pencil of quadrics of the type defined by two conics with two common points.

9. Further properties of quadrics.* Let S be a given quadric which meets a plane π in a conic C (assumed non-degenerate), and let Ω be another non-degenerate conic in π. Denote by O the pole of π with respect to S. As the notation implies, we have in mind the case when π is the 'plane at infinity' and O the centre of the quadric.

(i) Our first purpose is to consider a system of parallel chords of the quadric, and we begin with those chords which pass through a given point A in π. The harmonic conjugates of A with respect to the points where these chords meet S lie in a plane, namely, the polar plane of A; this plane passes through O since A lies in π. Hence we have the interpretation:

The middle points of a system of chords of a quadric which are parallel to a given direction lie in a plane through the centre of the quadric. The plane is called the *diametral plane conjugate to the given direction.*

Further, the plane meets π in the polar line of A with respect to the conic C. Hence, conversely, *each diametral plane defines a direction to which it is conjugate.*

(ii) With the help of the last remark, we can establish the existence of sets of three *conjugate diameters* of the quadric, with the property that the chords parallel to any one of them are bisected by the plane containing the other two. In fact, if PQR is any triangle lying in π and self-conjugate with respect to C, then the chords parallel to OP are bisected by the plane through O and the polar line of P with respect to C, i.e. by the plane OQR. Similar results hold for OQ and OR, as required.

(iii) Finally, we establish the existence, in the general case, of a set of three conjugate diameters which are also mutually perpendicular. These diameters are called the *principal axes* or,

* We shall now often quote results upon the 'plane at infinity' directly, assuming that the reader will fill in the details of interpretation for himself.

simply, the *axes* of the quadric, and the planes containing them are called the *principal planes*. Their existence is an immediate consequence of the theorem of plane geometry that, for general positions of the conics C and Ω in the plane π, they have a unique common self-polar triangle.

10. The circular sections; quadrics of revolution. With the notation of § 9, suppose that the conics C, Ω meet in four distinct points (the general case). Then any plane through two of these points cuts the quadric S in a conic through the two circular points at infinity in the plane, i.e. in a circle. Hence *there are six systems of parallel planes which cut S in circles* (real or 'imaginary'). The twelve points of the quadric, the tangent planes at which pass through two of the points of intersection of C, Ω, are called the *umbilics*; in other words, the umbilics are the points of the quadric such that plane sections parallel to the tangents there are circular.

Consider next the case when the conics C, Ω have double contact; let l be the chord of contact and L its pole with respect to the conics. Then OL is a diameter of the quadric S with the property that all planes perpendicular to it (passing, that is, through l) cut the quadric in circles with their centres on OL. The quadric S is therefore a *quadric of revolution* with OL as the *axis of revolution*.

A quadric of revolution can be regarded as having an infinite number of principal axes, namely OL and any two perpendicular diameters in the diametral plane perpendicular to OL.

11. Confocal quadrics and normals to quadrics; preliminary projective properties. The theories of confocal quadrics and of normals to quadrics turn out to be intimately connected, the link being found in the properties of tangential pencils. These properties have already been sketched as dual to those of ordinary pencils, but it is convenient at this stage to examine tangential pencils in their own right.

Consider, then, the two quadrics whose tangential equations are

$$\Sigma \equiv al^2+bm^2+cn^2+dp^2 = 0,$$

$$\Sigma' \equiv a'l^2+b'm^2+c'n^2+d'p^2 = 0.$$

For the particular metrical interpretation, we shall later take Σ' as the absolute conic, considered as a disc quadric, given by

$$a' = b' = c' = 1, \quad d' = 0.$$

The quadrics of the tangential pencil are defined by the equation

$$(a+\lambda a')\,l^2 + (b+\lambda b')\,m^2 + (c+\lambda c')\,n^2 + (d+\lambda d')\,p^2 = 0.$$

We assume, where necessary, that no two of a, b, c, d are proportional to the corresponding two of a', b', c', d'.

(i) *A unique quadric of the pencil touches a given arbitrary plane.* For if l, m, n, p are given, λ is determined.

(ii) *Three quadrics of the system pass through a given arbitrary point.*

For the equation of a quadric of the system in point coordinates is

$$\frac{x^2}{a+\lambda a'} + \frac{y^2}{b+\lambda b'} + \frac{z^2}{c+\lambda c'} + \frac{t^2}{d+\lambda d'} = 0,$$

and this equation, on clearing of fractions and regarding x, y, z, t as given, is cubic in λ.

(iii) *The tangent planes of two quadrics of a tangential pencil at a common point are conjugate with respect to all the quadrics of the pencil.*

The two planes are conjugate with respect to each of the two given quadrics, since the pole of either plane (being the point of contact with the relevant quadric) lies on the other. They are therefore, by the dual of Chapter VI, § 2 (iv), p. 100, conjugate with respect to all the quadrics of the tangential pencil.

(iv) *The poles of a given plane with respect to the quadrics of the system lie on a fixed line.*

For the pole of the plane whose coordinates are (L, M, N, P) is given by the equation

$$(a+\lambda a')\,Ll + (b+\lambda b')\,Mm + (c+\lambda c')\,Nn + (d+\lambda d')\,Pp = 0,$$

which, as λ varies, determines the points of a fixed line.

We call such a line an *axis* of the tangential pencil. An axis is thus a line for which a plane exists whose poles lie upon that line; we refer to the plane as *associated* to the axis. An axis meets its associated plane (assuming that it does not lie entirely in it; see (v) below) in the point of contact with that quadric of the pencil which touches the plane.

(v) *The associated plane of an axis which lies in it is a tangent plane of all the quadrics of the tangential pencil.*

For if the point determined by the equation in (iv) lies in the plane (L, M, N, P) for all values of λ, then

$$(a+\lambda a')L.L+(b+\lambda b')M.M+(c+\lambda c')N.N+(d+\lambda d')P.P\equiv 0;$$

the plane therefore touches the quadric whose equation is

$$(a+\lambda a')l^2+(b+\lambda b')m^2+(c+\lambda c')n^2+(d+\lambda d')p^2=0.$$

(vi) *Each axis belongs to a certain tetrahedral complex.*

For the line-coordinates of the line joining the points

$$(aL,\ bM,\ cN,\ dP),\quad (a'L,\ b'M,\ c'N,\ d'P)$$

are given by the expressions

$$l=(ad'-a'd)LP,\quad m=(bd'-b'd)MP,\quad n=(cd'-c'd)NP,$$
$$l'=(bc'-b'c)MN,\quad m'=(ca'-c'a)NL,\quad n'=(ab'-a'b)LM,$$

so that

$$\frac{ll'}{(bc'-b'c)(ad'-a'd)}=\frac{mm'}{(ca'-c'a)(bd'-b'd)}=\frac{nn'}{(ab'-a'b)(cd'-c'd)},$$

and the lines therefore belong to a tetrahedral complex.

Our hypothesis that no two of a, b, c, d are proportional to the corresponding two of a', b', c', d' ensures that the denominators do not vanish.

(vii) *Conversely, each line of the tetrahedral complex just defined is an axis of the tangential pencil.*

For suppose that l, m, n, l', m', n' are *given* numbers satisfying the relations

$$\frac{ll'}{(bc'-b'c)(ad'-a'd)}=\frac{mm'}{(ca'-c'a)(bd'-b'd)}=\frac{nn'}{(ab'-a'b)(cd'-c'd)},$$

and denote each of these three equal ratios by the letter u. We define the ratios of three numbers A, B, C by means of the relations

$$\frac{Al'}{bc'-b'c}=\frac{Bm'}{ca'-c'a}=\frac{Cn'}{ab'-a'b},$$

and we denote each of these three equal ratios by the letter v. Finally, we write

$$D = \rho u/v,$$

where ρ will be given a precise value later.

Then, as in (vi), the axis of the plane (A, B, C, D) has line-coordinates $(l_1, m_1, n_1, l_1', m_1', n_1')$, where

$$l_1 = (ad'-a'd)\,AD = \left\{\frac{ll'}{(bc'-b'c)\,u}\right\}\left\{\frac{(bc'-b'c)\,v}{l'}\right\}\left\{\frac{\rho u}{v}\right\} = \rho l,$$

$$l_1' = (bc'-b'c)\,BC = \frac{ABCl'}{v},$$

etc., and it follows that, if we select ρ to have the value ABC/v, then the plane (A, B, C, D) has the given line as axis.

The equation of the plane $Ax+By+Cz+Dt = 0$ can be expressed in terms of the given numbers $a, \ldots, a', \ldots, l, \ldots, l' \ldots$ (implying the number u defined above) in the form

$$\frac{m'n'x}{(ca'-c'a)(ab'-a'b)} + \frac{n'l'y}{(ab'-a'b)(bc'-b'c)} + \frac{l'm'z}{(bc'-b'c)(ca'-c'a)} + ut = 0.$$

Corollary. *The axes which pass through an arbitrary point generate a quadric cone and the axes which lie in an arbitrary plane touch a conic.*

(viii) *The polar planes of a given point with respect to the quadrics of a tangential pencil belong to a cubic developable* [the dual of a twisted cubic curve].

For the polar plane of the point (ξ, η, ζ, τ) with respect to the quadric

$$\frac{x^2}{a+\lambda a'} + \frac{y^2}{b+\lambda b'} + \frac{z^2}{c+\lambda c'} + \frac{t^2}{d+\lambda d'} = 0$$

is

$$\frac{\xi x}{a+\lambda a'} + \frac{\eta y}{b+\lambda b'} + \frac{\zeta z}{c+\lambda c'} + \frac{\tau t}{d+\lambda d'} = 0,$$

and the coordinates of the plane are therefore expressible in terms of the parameter λ in the form

$$\left(\frac{\xi}{a+\lambda a'}, \frac{\eta}{b+\lambda b'}, \frac{\zeta}{c+\lambda c'}, \frac{\tau}{d+\lambda d'}\right).$$

On multiplying by the common denominator

$$(a+\lambda a')(b+\lambda b')(c+\lambda c')(d+\lambda d'),$$

the coordinates are expressed as cubic polynomials in the parameter; the planes therefore belong to the dual of a cubic curve, that is, to a cubic developable.

Thus three planes of the system pass through the arbitrary point (x_1, y_1, z_1, t_1) whose equation is $lx_1 + my_1 + nz_1 + pt_1 = 0$.

When λ takes the values $-a/a'$, $-b/b'$, $-c/c'$, $-d/d'$, we obtain respectively the planes whose coordinates are

$$(1, 0, 0, 0), \quad (0, 1, 0, 0), \quad (0, 0, 1, 0), \quad (0, 0, 0, 1);$$

hence *the faces of the common self-polar tetrahedron belong to the developable, whatever the position of the given point* (ξ, η, ζ, τ).

(ix) *Six axes can be drawn through an arbitrary given point so that their associated planes touch any given quadric of the tangential pencil.*

For the polar planes of a given point with respect to all the quadrics of the pencil belong to a cubic developable. We have to pick out those particular planes of the cubic developable which also belong to the given quadric regarded as an envelope of planes; for the pole of such a plane with respect to the given quadric is the given point, which therefore lies on the associated axis. Now a cubic developable and a quadric envelope have *six* planes in common, which is the required result, since each plane determines an axis.

The reader may at first find the dual statement more convincing: a twisted cubic and a quadric surface have six points in common.

(x) *The twisted cubic, through the six points of contact of those planes of the cubic developable* [*defined as in* (viii) *from a given point* P] *which touch a given quadric of the tangential pencil, passes through* P *and the four vertices of the common self-polar tetrahedron of the quadrics.*

Reciprocate the figure with respect to the given quadric, which thus reciprocates into itself. The planes of the cubic developable become the points of a twisted cubic curve Γ; in particular, the six planes common to the developable and the quadric envelope become the six points of contact, so that Γ is the twisted cubic described in the enunciation. Also the developable, by definition, contains the polar plane of P with respect to the given quadric, and the reciprocal of this plane is P itself. Finally, the developable, as we proved, contains the four faces of the common self-polar tetrahedron; these faces reciprocate into the four vertices, which therefore lie on the twisted cubic.

12. Confocal quadrics and normals to quadrics; metrical properties. After the somewhat lengthy digression of the last paragraph, we return to our main theme. A (general) tangential pencil contains four disc quadrics, of which we now select one as the absolute conic. We can, without loss of generality, take this conic as one of the quadrics defining the pencil, and also transform its equation to have unit coefficients. Thus we take the equation of the absolute conic in the form

$$l^2+m^2+n^2 = 0,$$

and the equation of the pencil is then

$$(a+\lambda)\,l^2+(b+\lambda)\,m^2+(c+\lambda)\,n^2+dp^2 = 0.$$

We call the quadrics so defined a system of *confocal central quadrics*,* the centre (with $t = 0$ as the plane at infinity) being the point $(0, 0, 0, 1)$.

More generally, if $\Sigma = 0$ is the tangential equation of any quadric, then the quadrics

$$\Sigma+\lambda(l^2+m^2+n^2) = 0,$$

where $l^2+m^2+n^2 = 0$ is the absolute conic, are said to form a confocal system. We restrict ourselves to the special case just defined.

We can now interpret several of the results of § 11 metrically, as follows:

(i) A unique quadric of a confocal system touches a given arbitrary plane.

(ii) Three quadrics of a confocal system pass through an arbitrary point.

* The analogy with confocal conics is obvious. We shall not, however, be considering the properties of the 'foci' of quadrics.

(iii) Two quadrics of a confocal system cut at right angles. [Remember, in interpreting § 11 (iii), that the absolute conic is one of the quadrics of the system.] Thus the three quadrics defined in (ii) have the property that their tangent planes at the given point are mutually perpendicular.

(iv), (v). Suppose now that S is a given quadric of the confocal system and P a given point upon it. Let ρ be the tangent plane to S at P. Then the poles of ρ with respect to the individual quadrics of the system all lie on a certain straight line through P which (since the absolute conic is one of the quadrics) is perpendicular to the plane ρ. This line, which is an axis of the tangential pencil, is called the *normal* to S at P.

The case when the axis lies in its associated plane arises when the plane-coordinates of that plane satisfy the relation $l^2+m^2+n^2 = 0$, so that such axes (in ordinary Euclidean geometry) are not real.

We now obtain properties of the normals of quadrics from the results of § 11, as follows:

(vi), (vii). The normals of the quadrics of a confocal system all belong to a certain tetrahedral complex; and, conversely, the lines of that complex are normals of such quadrics.

(viii). The normals which pass through a given point lie on a certain quadric cone, and the normals which lie in a given plane touch a certain conic.

(ix), (x). Six normals to a given quadric can be drawn through an arbitrary point. The twisted cubic through the feet of these normals passes through the given point and the centre of the quadric, and the cubic has an asymptote parallel to each principal axis of the quadric.

Finally, the confocal system contains three disc quadrics other than the absolute conic. They are called the *focal conics*.

Other properties of confocal quadrics and normals are suggested in the examples which follow, and the reader should work them out carefully. A full and detailed account is given in Professor Baker's *Principles of Geometry*, Vol. III, Chapter 2, where other metrical results may also be found.

MISCELLANEOUS EXAMPLES VII

1.* Prove that, if two planes are each perpendicular to a third plane, their line of intersection is perpendicular to that plane.

Prove that, if the perpendiculars from two vertices of a tetrahedron on the opposite faces intersect, then the perpendiculars from the other two vertices on the opposite faces also intersect. [M.T. I.]

2. In a quadrilateral $ABCD$, which does not lie in a plane, prove that the middle points of the four sides lie in a plane and determine a parallelogram. [M.T. I.]

3. From a point O perpendiculars Oa, Ob, Oc, Od are drawn on the faces of a tetrahedron $ABCD$. Prove that the pairs of lines such as (AB, cd), (BC, ad) are mutually perpendicular.

Hence prove that any pair of perpendiculars from A, B, C, D on the corresponding faces of the tetrahedron $abcd$ are coplanar and then prove that all these perpendiculars are concurrent. [M.T. I.]

4. Prove that, if a sphere cuts two given spheres orthogonally, it will also cut orthogonally every sphere of the radical system determined by them. [M.T. I (adapted).]

5. Two circles, in different planes, meet in the two points A, B. Show that a sphere S can be drawn to pass through both circles.

If the circles intersect at right angles (so that their tangents at A and at B are perpendicular), prove that the plane of either circle passes through the pole of the plane of the other with respect to the sphere S. [M.T. I.]

6. A variable quadric has the edges AB, BC, CD of a tetrahedron $ABCD$ as generators. Show that the locus of the pole of the plane bisecting the edges AB, CD, AC, BD is the plane bisecting AB, CD, AD, BC. [L.]

7. If three given non-intersecting lines are parallel to a plane, show that this is also true of the lines which meet them. [M.T. II.]

8. Prove that, if the perpendicular lines λ, λ' are polar lines with respect to a quadric S, then each of λ, λ' is a principal axis of some plane section of S. [M.T. II.]

9. If the tangent cone from a point P to a central quadric is circular, prove that P lies on one of three conics, and state (without proof) the corresponding result for a paraboloid.

Prove also that the axis of the circular tangent cone is the tangent at P to the conic-locus of P. [P.]

10. A variable line l through a given point is perpendicular to its polar line with respect to a given quadric. Prove that l generates a quadric cone.

* It is obvious that several of these examples were not expected to be answered by the methods described in this chapter, but the reader may like scope for practice.

11. A quadric is described having two given skew lines as generators and passing through a given point. Prove that its centre lies in a fixed plane. [P.]

12. Show that a quadric of revolution cuts the plane at infinity in a conic having double contact with the imaginary circle at infinity (the absolute conic).

Prove that the axes of revolution of quadrics of revolution containing two fixed non-intersecting straight lines in space are always perpendicular to one or other of two fixed directions. [M.T. II.]

13. $\alpha, \lambda, \mu, \nu$ are four concurrent planes. On the lines $(\alpha\lambda)$, $(\alpha\mu)$, $(\alpha\nu)$ are taken points L, M, N respectively, and planes are drawn through MN, NL, LM parallel respectively to $(\mu\nu)$, $(\nu\lambda)$, $(\lambda\mu)$. Show that each of these planes is parallel to the line of intersection of the other two. [M.T. II.]

14. Prove that the plane sections of a quadric are projected from a point O of the quadric into conics passing through two fixed points in the plane ϖ of projection. If these fixed points are interpreted as the circular points in ϖ, show that (i) the centre of a circle in ϖ is the intersection with ϖ of the line joining O to the pole of the plane of the corresponding section of the quadric; (ii) orthogonal circles arise from conjugate planes.

If S, S_1, S_2 are three conics on a quadric such that S touches S_1 and S_2 in P and Q respectively, show that PQ passes through the vertex of one of the cones through S_1 and S_2. Hence show that there are eight circles which touch three given circles. [M.T. II.]

15. Explain how propositions on circles in a plane can be stated as propositions on plane sections of a quadric.

Establish the equivalence of the two theorems:

(i) The circumcircles of the four triangles obtained by omitting in turn each one of four arbitrary lines in a plane meet in a point.

(ii) Through a point O of a quadric four arbitrary planes α_1, α_2, α_3, α_4 are drawn, and the line $\alpha_i\alpha_j$ meets the quadric again in P_{ij}. The four planes $P_{23}P_{31}P_{12}$, $P_{34}P_{42}P_{23}$, etc. meet in a point O' of the quadric.

Prove one of these theorems [the reader will naturally choose the second] and show that the eight points O, O', P_{ij} are such that quadrics through any seven pass through the eighth. [M.T. II.]

16. In one system of generators of a particular quadric there are three generators at right angles to one another. Prove that there is an infinite number of such mutually orthogonal triads of generators in the system.

Prove also that if three generators of the system are taken at random, then there is a fourth generator of the system such that the common perpendicular of any two of the four is at right angles to the common perpendicular of the other two. [M.T. II.]

17. A quadric and a sphere intersect the plane at infinity in four common points. Prove the following properties of these four points:

(i) Any circular section of the quadric passes through two of the four points.

(ii) The twelve umbilics (real or imaginary) of the quadric lie three by three on the eight generators of the quadric which pass through the four points.

Show that through any point six planes can be drawn so as to have the point as a focus of the sections of a given quadric by the planes. [M.T. II.]

18. Prove that the perpendiculars from B, C, D to the opposite faces of the tetrahedron $ABCD$ meet the perpendicular to the plane of the triangle BCD through the orthocentre of this triangle. Hence show that the perpendiculars from the vertices of a tetrahedron to the opposite faces lie on a quadric Ω, and that the perpendiculars to the faces of the tetrahedron through the orthocentres also lie on Ω.

Show that each face of the tetrahedron meets Ω in a rectangular hyperbola, and deduce that Ω contains a set of three mutually perpendicular generators. [M.T. II.]

19. Prove that the locus of the poles of a given plane ϖ with regard to the quadrics of a confocal system is a straight line l, which is a normal to that one of the quadrics which touches ϖ.

Prove further that the normals to the quadrics which lie in the plane ϖ touch a parabola and are the polar lines of the line l with regard to the individual quadrics of the system. Hence or otherwise prove that, if the normals at P, Q to a quadric S are concurrent, then PQ is a normal to a confocal quadric. [P.]

20. (i) Prove that those points of a central quadric at which the normals intersect the normal at a fixed point P lie on a quadric cone whose vertex is P. Hence, or otherwise, show that each of the normals at the points of a plane section S is met by the normals at two other points of S.

(ii) Prove that, of all the normals to the quadric at the points of a plane section S, there are two which lie in the plane of S, giving a geometrical construction for them. [P.]

21. Prove that six normals can be drawn from a point to a central quadric, and that these normals are generators of a quadric cone, each of whose generators is perpendicular to its polar line with respect to the quadric. [M.T. II.]

22. A line meets a quadric in the points P, Q. Prove that the normals at P, Q will intersect if PQ is perpendicular to its polar line, and hence or otherwise show that the lines which lie in a given plane and which meet the quadric in two points, the normals at which intersect, touch a parabola. [M.T. II.]

23. Prove that those normals to the quadrics of a confocal system which lie in a given plane π envelope a parabola touching the three principal planes.

Show that to each quadric there are two normals lying in π, and that the locus of their intersection as the quadric varies in the confocal system is a line. [M.T. II.]

24. Prove that the polar lines of a given line with regard to the quadrics of a confocal system are in general generators of a quadric, but that if the line is a normal to one of the confocals the quadric degenerates (tangentially) into a conic.

Prove also that if a line is perpendicular to its polar line with regard to a quadric it is a normal to a confocal quadric, and that if the normals to a quadric at two points P and Q lie in a plane then PQ is a normal to a confocal. [M.T. II.]

25. Prove that the normals drawn from a point P to the quadrics of a confocal system generate a quadric cone passing through the centre of the quadrics and containing a generator parallel to each of the principal axes.

Show that six normals can be drawn from P to a general quadric of the system, and that their feet lie on a twisted cubic passing through P and the centre of the quadric and having its asymptotes parallel to the axes. [M.T. II.]

26. Prove that all the planes which touch a given central quadric Σ and the absolute conic Ω touch each of three other fixed conics f_1, f_2, f_3. If P is any point of one of f_1, f_2, f_3, prove that the tangent cone from P to Σ is right circular.

Deduce that six right circular cylinders [not necessarily real] can be drawn to envelop Σ. [L.]

27. Prove that the tangential equation of the disc quadric whose boundary is the section of the quadric by a plane ω is

$$\Sigma\Sigma_{\omega\omega} - \Sigma_\omega^2 = 0.$$

If the disc quadric $\Omega = 0$ belongs to the range determined by the two quadrics $\Sigma = 0$, $\Phi = 0$, prove that the common tangent planes of Φ and the section of Σ by the double plane of Ω touch a quadric through the boundary of Ω.

Show that the planes of parabolic section of a central quadric Σ which touch a focal conic of Σ all touch a sphere concentric with Σ. [L.]

28. Explain what is meant by the 'absolute conic' ('circle at infinity'), and show that the generators of a sphere all meet the absolute conic.

A conic and a circle which touches it at two points are rotated about the axis of the conic on which the centre of the circle lies, giving a sphere and a quadric of revolution which touch at all points of a circle. Show that the tangent plane at a point S of the sphere intersects the quadric in a conic which has S as a focus. [L.]

CHAPTER VIII

THE USE OF MATRICES

It is not an easy matter for the writer of a text-book at the present level to judge the moment when he should introduce matrices, or, having introduced them, to decide how far he should go. We have taken the line that in an introductory course the reader should first become familiar with the ideas and methods as they arise in direct development from plane geometry, and have therefore kept to the notation x, y, z, t for points, l, m, n, p for planes, and l, m, n, l', m', n' for lines.

By now, however, the fundamental ideas should be established, and the reader ready for the more unified treatment which the algebra of matrices affords. We therefore conclude with a chapter in which the first section gives a brief but, we hope, fairly complete account of the algebra as required in geometry, and the second shows how to apply these new ideas to what should be old and familiar ground. The account is brief, and its purpose introductory. The reader is strongly urged to follow it up, for example in the book which is likely to become a classic, Dr J. A. Todd's *Projective and Analytical Geometry* (Pitman, 1947).

SECTION I. MATRICES

1. First properties. DEFINITION. A set of mn elements

$$\begin{matrix} a_{11} & a_{12} & a_{13} & \dots & a_{1n} \\ a_{21} & a_{22} & a_{23} & \dots & a_{2n} \\ \dots & \dots & \dots & \dots & \dots \\ a_{m1} & a_{m2} & a_{m3} & \dots & a_{mn} \end{matrix}$$

organised into m rows and n columns for operations to be defined below, is called a *matrix*, of type $m \times n$. A typical element is a_{ij}, where the suffix i refers to the row and j to the column. The matrix as a whole is denoted briefly by the symbol $\mathbf{a}$, or by the notation (a_{ij}).

When $m = n$, the matrix is called *square*. If, in addition, the elements a_{ij}, a_{ji} are equal, the matrix is *symmetrical*; and, if the elements a_{ij}, a_{ji} are equal and opposite, the matrix is *skew-symmetrical*. Thus the matrices

$$\begin{pmatrix} a & h & g \\ h & b & f \\ g & f & c \end{pmatrix}, \quad \begin{pmatrix} 0 & -n & m \\ n & 0 & -l \\ -m & l & 0 \end{pmatrix}$$

are respectively symmetrical and skew-symmetrical.

The 'leading elements' a_{ii} of a skew-symmetrical matrix are all zero.

When all of the elements of any matrix are zero, the matrix itself is called a *zero* matrix, denoted by $\mathbf{0}$.

An $n \times n$ matrix such that $a_{ii} = 1$ and $a_{ij} = 0\,(i \neq j)$ is called the *unit* matrix $\mathbf{I}_n$ of n rows and columns. For example,

$$\mathbf{I}_3 \equiv \begin{pmatrix} 1 & 0 & 0 \\ 0 & 1 & 0 \\ 0 & 0 & 1 \end{pmatrix}.$$

When a matrix consists of a single column of m rows, it is called a *column vector*; a single row of n columns is called a *row vector*. For example,

$$\begin{pmatrix} a \\ b \\ c \end{pmatrix}, \quad (a, b, c)$$

are respectively a column vector and a row vector; but it is more convenient to use the 'horizontal' notation

$$\{a, b, c\}$$

for a column vector.

An 'ordinary' number of complex algebra, when used in this context, is called a *scalar*.

A letter such as $\mathbf{a}$, in bold type, will denote a matrix or vector. A letter such as a, in *italic*, will denote a scalar.

The determinant $|a_{ij}|$ whose elements are those of a square matrix $\mathbf{a}$ is called the *determinant of the matrix*, and is also denoted by the symbol $|\mathbf{a}|$, or det $\mathbf{a}$.

2. Multiplication by a scalar. The *product* of a matrix $\mathbf{a}$ by a scalar p is defined to be the matrix $\mathbf{b}$ each of whose elements is p times the corresponding element of $\mathbf{a}$. Thus

$$b_{ij} = pa_{ij}.$$

For example,

$$2\begin{pmatrix} a & h & g \\ h & b & f \\ g & f & c \end{pmatrix} = \begin{pmatrix} 2a & 2h & 2g \\ 2h & 2b & 2f \\ 2g & 2f & 2c \end{pmatrix}.$$

3. Scalar linear combinations of matrices. If $\mathbf{a}$, $\mathbf{b}$ are two matrices and p, q two scalars, then the expression

$$p\mathbf{a} + q\mathbf{b}$$

is defined to be the matrix, conveniently denoted by $\mathbf{c}$, such that

$$c_{ij} = pa_{ij} + qb_{ij}.$$

This is possible, of course, only if $\mathbf{a}$, $\mathbf{b}$ are matrices of the same type, each having (say) m rows and n columns.

In particular, $p = q = 1$ gives the *sum* $\mathbf{a}+\mathbf{b}$, and $p = 1$, $q = -1$ gives the *difference* $\mathbf{a}-\mathbf{b}$.

For example, if

$$\mathbf{a} \equiv \begin{pmatrix} l_1 & m_1 & n_1 \\ l_2 & m_2 & n_2 \end{pmatrix}, \quad \mathbf{b} \equiv \begin{pmatrix} \lambda_1 & \mu_1 & \nu_1 \\ \lambda_2 & \mu_2 & \nu_2 \end{pmatrix},$$

then

$$\mathbf{a}-\mathbf{b} = \begin{pmatrix} l_1-\lambda_1 & m_1-\mu_1 & n_1-\nu_1 \\ l_2-\lambda_2 & m_2-\mu_2 & n_2-\nu_2 \end{pmatrix}.$$

4. Matrix multiplication. Definition. The *product* of two matrices $\mathbf{a}$, $\mathbf{b}$ is defined to be that matrix, conveniently denoted by $\mathbf{c}$, such that

$$c_{ij} = \sum_{\lambda} a_{i\lambda} b_{\lambda j},$$

the summation being for $\lambda = 1, 2, 3, \ldots$. But this requires more explanation.

In the first place, λ refers to columns of **a** and to rows of **b**, and so the summation is possible only if the number of columns of **a** is equal to the number of rows of **b**. Thus if **a** is of type $m \times n$, then **b** must be $n \times p$. Two such matrices are said to be *conformable for multiplication in the order* **ab**.

This leads to the second point, that multiplication in the order **ab** is by no means the same as multiplication in the order **ba**. The latter is, in fact, the matrix **d**, where

$$d_{ij} = \sum_{\mu} b_{i\mu} a_{\mu j}.$$

The existence of **ba** of course requires the relation $m = p$.

For example, if

$$\mathbf{a} \equiv \begin{pmatrix} l_1 & l_2 \\ m_1 & m_2 \\ n_1 & n_2 \end{pmatrix}, \qquad \mathbf{b} \equiv \begin{pmatrix} x_1 & x_2 & x_3 \\ y_1 & y_2 & y_3 \end{pmatrix},$$

then

$$\mathbf{ab} = \begin{pmatrix} l_1x_1 + l_2y_1 & l_1x_2 + l_2y_2 & l_1x_3 + l_2y_3 \\ m_1x_1 + m_2y_1 & m_1x_2 + m_2y_2 & m_1x_3 + m_2y_3 \\ n_1x_1 + n_2y_1 & n_1x_2 + n_2y_2 & n_1x_3 + n_2y_3 \end{pmatrix}$$

and

$$\mathbf{ba} = \begin{pmatrix} x_1l_1 + x_2m_1 + x_3n_1 & x_1l_2 + x_2m_2 + x_3n_2 \\ y_1l_1 + y_2m_1 + y_3n_1 & y_1l_2 + y_2m_2 + y_3n_2 \end{pmatrix}.$$

Finally, since, in the definition

$$c_{ij} = \sum_{\lambda} a_{i\lambda} b_{\lambda j},$$

i refers to rows of **a** while j refers to columns of **b**, the matrix **c** is derived from m values of i and p of j, so that it is a matrix of type $m \times p$. This gives the 'domino' rule, that the product of two matrices of type $m \times n$, $n \times p$ is (for that order of multiplication) of type $m \times p$.

By repeated application of these rules, we may clearly form further products **abc**, **abcd**, and so on. Note that these products obey the associative law of multiplication.

Notation. When we come to applications in geometry, we shall find a number of occasions when a $1 \times n$ matrix **a** is multiplied by an $n \times 1$ matrix **b**. The result is a 1×1 matrix, that is, a scalar **ab**.

It will be convenient to denote the product when it is a scalar by the symbol (**ab**) in brackets. Such a scalar has all the ordinary properties of complex numbers; in particular, the product of a matrix **c** by the scalar (**ab**) is the matrix, equally well denoted by the symbols

$$(\mathbf{ab})\,\mathbf{c}, \quad \mathbf{c}(\mathbf{ab}),$$

whose elements are (**ab**) times the corresponding elements of **c**.

We add a point of manipulation. If, say, **a**, **b**, **c**, **d** are matrices of types $m \times 1$, $1 \times p$, $p \times 1$, $1 \times n$, then the product **abcd** contains the scalar (**bc**) as a factor, and this may be 'taken out' like an 'ordinary' number. For example, the product is equal to (**bc**) **ad**.

Finally, it is easy to verify that

$$\mathbf{aI}_n = \mathbf{a}, \quad \mathbf{I}_m\mathbf{a} = \mathbf{a},$$

corresponding to the standard property of the unit of ordinary algebra.

5. The transpose of a matrix. Definition. The *transpose* of a matrix **a** is the matrix, denoted by the symbol **a**′, whose elements are obtained from those of **a** by interchanging rows and columns. Thus

$$a'_{ij} = a_{ji}.$$

Then **a**′ is a matrix of type $n \times m$.

For example, if

$$\mathbf{a} \equiv \begin{pmatrix} l_1 & m_1 & n_1 \\ l_2 & m_2 & n_2 \end{pmatrix},$$

then

$$\mathbf{a}' \equiv \begin{pmatrix} l_1 & l_2 \\ m_1 & m_2 \\ n_1 & n_2 \end{pmatrix}.$$

A *symmetrical* matrix is characterised by the relation $\mathbf{a}' = \mathbf{a}$, and a *skew-symmetrical* matrix by the relation $\mathbf{a}' = -\mathbf{a}$.

We prove that, *if* **a**, **b** *are two given matrices, then the transpose* [**ab**]′ *of the product* **ab** *is given by* **b**′**a**′, *in that order.*

Consider the product $\mathbf{b}'\mathbf{a}'$. We have

$$\begin{aligned}[\mathbf{b}'\mathbf{a}']_{ij} &= \sum_{\lambda} b'_{i\lambda} a'_{\lambda j} \\ &= \sum_{\lambda} b_{\lambda i} a_{j\lambda} \\ &= \sum_{\lambda} a_{j\lambda} b_{\lambda i} \\ &= [\mathbf{ab}]'_{ij}.\end{aligned}$$

6. The inverse matrix. Let $\mathbf{a}$ be a given matrix. We seek to define, if possible, a matrix, conveniently denoted by the notation $\mathbf{a}^{-1}$, with the property that both of the products $\mathbf{aa}^{-1}$ and $\mathbf{a}^{-1}\mathbf{a}$ exist and are equal to the unit matrix $\mathbf{I}_n$. Since the product $\mathbf{aa}^{-1}$ is $\mathbf{I}_n$, the matrix $\mathbf{a}$ must have n rows; and since $\mathbf{a}^{-1}\mathbf{a}$ is also $\mathbf{I}_n$, it must have n columns, so that $\mathbf{a}$ *is a square* $n \times n$ *matrix.*

The matrix $\mathbf{a}^{-1}$, when it exists, is called the *inverse* of $\mathbf{a}$.

In order to identify $\mathbf{a}^{-1}$, we appeal to the well-known theorem on determinants, used in earlier chapters, that, if α_{ij} is the cofactor of the element a_{ij} in the determinant $|\mathbf{a}|$, then

$$\sum_{\lambda=1} a_{i\lambda}\alpha_{j\lambda} = \begin{cases} |\mathbf{a}| & \text{if} \quad i = j, \\ 0 & \text{if} \quad i \neq j. \end{cases}$$

Now denote by $\mathbf{A}$ the TRANSPOSE of the matrix $\boldsymbol{\alpha}$, so that

$$A_{\lambda j} = \alpha_{j\lambda}.$$

Then the result just quoted may be written in matrix notation in the form

$$\mathbf{aA} = |\mathbf{a}|\,\mathbf{I}_n.$$

Similarly, the relation

$$\sum_{\mu=1}^{n} a_{\mu i}\alpha_{\mu j} = \begin{cases} |\mathbf{a}| & \text{if} \quad i = j, \\ 0 & \text{if} \quad i \neq j \end{cases}$$

leads to the matrix equation

$$\mathbf{Aa} = |\mathbf{a}|\,\mathbf{I}_n.$$

The matrix $\mathbf{a}^{-1}$ is therefore identified by the relation

$$\mathbf{a}^{-1} = \frac{\mathbf{A}}{|\mathbf{a}|}.$$

DEFINITION. The matrix $\mathbf{A}$, which is the *transpose* of the matrix of cofactors of $\mathbf{a}$, is called the *adjoint* (or *adjugate*) of $\mathbf{a}$, sometimes written

$$\text{adj. } \mathbf{a}.$$

Note the relation

$$\mathbf{a} \,\text{adj.}\, \mathbf{a} = |\mathbf{a}|\, \mathbf{I}_n = \text{adj. } \mathbf{a}.\mathbf{a},$$

or

$$\mathbf{aA} = |\mathbf{a}|\, \mathbf{I}_n = \mathbf{Aa}.$$

We conclude with one or two important properties.

I. *To prove that* $\quad |\mathbf{A}| = |\mathbf{a}|^{n-1}.$

LEMMAS. (i) *If* $\mathbf{c} = \mathbf{ab}$, *where* $\mathbf{a}, \mathbf{b}$ *are square matrices, then*

$$|\mathbf{c}| = |\mathbf{a}|.|\mathbf{b}|.$$

This is merely the restatement of one of the standard rules for the multiplication of determinants.

(ii) *If* $\mathbf{c} = p\mathbf{a}$, *where* p *is a scalar, then*

$$|\mathbf{c}| = p^n |\mathbf{a}|.$$

For each element of $\mathbf{a}$ is multiplied by p, and so each of the n rows (or columns) of $|\mathbf{c}|$ contains p as a factor.

The proof of the main theorem is immediate:

Since $\quad \mathbf{aA} = |\mathbf{a}|.\mathbf{I}_n,$

and the determinant $|\mathbf{a}|$ is a scalar, we have

$$|\mathbf{a}|.|\mathbf{A}| = |\mathbf{a}|^n |\mathbf{I}_n| = |\mathbf{a}|^n.$$

Hence $\quad |\mathbf{A}| = |\mathbf{a}|^{n-1}.$

COROLLARY. $\quad \mathbf{A} \,\text{adj.}\, \mathbf{A} = |\mathbf{A}|\, \mathbf{I}_n = |\mathbf{a}|^{n-1} \mathbf{I}_n$

$$= \text{adj.}\, \mathbf{A}.\mathbf{A}.$$

II. *If* $\mathbf{c} = \mathbf{ab}$, *then* $\mathbf{c}^{-1} = \mathbf{b}^{-1}\mathbf{a}^{-1}$.

Consider the product

$$\mathbf{abb}^{-1}\mathbf{a}^{-1} = \mathbf{aI}_n\mathbf{a}^{-1} = \mathbf{aa}^{-1}$$

$$= \mathbf{I}_n.$$

Hence $\mathbf{b}^{-1}\mathbf{a}^{-1}$ is, by definition, the inverse of $\mathbf{ab}$.

This result may be extended. For example,

$$[\mathbf{ab}^{-1}\mathbf{c}]^{-1} = \mathbf{c}^{-1}\mathbf{ba}^{-1}.$$

We have held back the final point to avoid breaking the argument, but it is of fundamental importance. Throughout our work we have implicitly assumed that the determinant $|\mathbf{a}|$ is not zero. In other words, *a matrix whose determinant is zero has no inverse.*

A matrix with zero determinant is said to be *singular*.

Section II. Geometrical Applications

7. Coordinates and the plane. As we set out to apply the theory of matrices, we rename our coordinates to take advantage of the suffix notation. In 'matrix geometry', as we may call it, there is often no need to specify exactly the dimensions of the space under discussion; equations are established which, suitably interpreted, are applicable equally to the plane, the three-dimensional space with which this book has been dealing, or any higher space. We assume, however, that our attention is confined to three dimensions, and we name the elements accordingly.

We require four homogeneous coordinates for a point; the standard notation is x_0, x_1, x_2, x_3 for the coordinates of a point X, arranging them as a column vector $\mathbf{x} \equiv \{x_0, x_1, x_2, x_3\}$. The use of the suffix 0 for the first coordinate is customary; it has the effect of replacing some of the summations of the matrix algebra with a lower limit 0 instead of 1, but the modifications are obvious. In making comparisons with previous work, the reader may like to identify (x_0, x_1, x_2, x_3) with (t, x, y, z) in that order; but this is not fundamentally important.

The coordinate vector of a point P on a line XY can be expressed by means of the relation

$$\mathbf{p} = \lambda\mathbf{x} + \mu\mathbf{y}$$

[meaning $\{p_0, p_1, p_2, p_3\} = \lambda\{x_0, x_1, x_2, x_3\} + \mu\{y_0, y_1, y_2, y_3\}$], and the vector of a point P on a plane XYZ by the relation

$$\mathbf{p} = \lambda\mathbf{x} + \mu\mathbf{y} + \nu\mathbf{z}.$$

The equation of a plane U will be taken in the form

$$u_0x_0 + u_1x_1 + u_2x_2 + u_3x_3 = 0,$$

so that $$(\mathbf{u}'\mathbf{x}) = 0, \quad \text{or} \quad (\mathbf{x}'\mathbf{u}) = 0,$$

where $(\mathbf{u}'\mathbf{x})$ is a scalar. The plane U is thus defined by its plane coordinates u_0, u_1, u_2, u_3 which form the elements of a column vector **u**.

Illustration 1. *To find the coordinates of the point where the line joining the points* **p**, **q** *meets the plane* **u**.

Any point of **p**, **q** is **r**, where

$$\mathbf{r} = \lambda\mathbf{p} + \mu\mathbf{q}.$$

If it lies in the plane **u**, then $(\mathbf{u}'\mathbf{r}) = 0$, so that

$$\lambda(\mathbf{u}'\mathbf{p}) + \mu(\mathbf{u}'\mathbf{q}) = 0,$$

or

$$\frac{\lambda}{(\mathbf{u}'\mathbf{q})} = \frac{\mu}{-(\mathbf{u}'\mathbf{p})}.$$

Absorbing a suitable multiplier into the coordinates of **r**, we obtain its vector in the form

$$(\mathbf{u}'\mathbf{q})\,\mathbf{p} - (\mathbf{u}'\mathbf{p})\,\mathbf{q}.$$

8. Transformation of coordinates. In dealing with transformations, we use an asterisk (*) to denote corresponding elements, as a dash (′) is now required to denote transposition of matrices.

The matrix equation

$$\mathbf{x} = \mathbf{T}\mathbf{x}^*,$$

where **T** is a non-singular square 4×4 matrix, defines a *transformation* between two coordinate systems, changing the coordinate vector **x** in one system into $\mathbf{x}^*$ in the other. In more detail, the coordinate x_i becomes $\sum_{\lambda=0}^{3} T_{i\lambda} x_\lambda^*$ for $i = 0, 1, 2, 3$.

To find the reverse of this transformation, multiply each side on the left by the inverse matrix $\mathbf{T}^{-1}$, giving

$$\mathbf{x}^* = \mathbf{T}^{-1}\mathbf{x}.$$

The corresponding transformation for planes can be found immediately. The equation of the plane defined by the vector **u** is

$$(\mathbf{u}'\mathbf{x}) = 0,$$

and this, on transformation, becomes

$$(\mathbf{u}'\mathbf{T}\mathbf{x}^*) = 0,$$

or

$$(\mathbf{u}^{*\prime}\mathbf{x}^*) = 0,$$

where

$$\mathbf{u}^{*\prime} = \mathbf{u}'\mathbf{T},$$

or

$$\mathbf{u}^* = \mathbf{T}'\mathbf{u}.$$

Hence *the transformation* $\mathbf{T}$ *acts according to the equations*

$$\mathbf{x} = \mathbf{T}\mathbf{x}^*, \quad \mathbf{x}^* = \mathbf{T}^{-1}\mathbf{x}$$

for point vectors, and

$$\mathbf{u} = (\mathbf{T}')^{-1}\mathbf{u}^*, \quad \mathbf{u}^* = \mathbf{T}'\mathbf{u}$$

for plane vectors.

9. The bilinear form. Consider the scalar or, as it is called, the *bilinear* form

$$s \equiv (\mathbf{x}'\mathbf{a}\mathbf{y}),$$

where $\mathbf{x}$, $\mathbf{y}$ are column vectors and $\mathbf{a}$ a 4×4 matrix. On expansion,

$$s \equiv x_0 \sum_{\lambda=0}^{3} a_{0\lambda} y_\lambda + x_1 \sum_{\lambda=0}^{3} a_{1\lambda} y_\lambda + x_2 \sum_{\lambda=0}^{3} a_{2\lambda} y_\lambda + x_3 \sum_{\lambda=0}^{3} a_{3\lambda} y_\lambda;$$

or, arranging in terms of y_i,

$$s \equiv y_0 \sum_{\mu=0}^{3} a_{\mu 0} x_\mu + y_1 \sum_{\mu=0}^{3} a_{\mu 1} x_\mu + y_2 \sum_{\mu=0}^{3} a_{\mu 2} x_\mu + y_3 \sum_{\mu=0}^{3} a_{\mu 3} x_\mu.$$

This is also the scalar form $(\mathbf{y}'\mathbf{a}'\mathbf{x})$,

so that

$$s \equiv (\mathbf{x}'\mathbf{a}\mathbf{y}) \equiv (\mathbf{y}'\mathbf{a}'\mathbf{x});$$

that is to say, the *bilinear form* $(\mathbf{x}'\mathbf{a}\mathbf{y})$ *is unaltered by transposition.*†

In particular, if $\mathbf{a}$ is symmetric, then $(\mathbf{x}'\mathbf{a}\mathbf{y}) = (\mathbf{y}'\mathbf{a}\mathbf{x})$.

Putting $\mathbf{y} = \mathbf{x}$, we obtain the *quadratic form*

$$q \equiv (\mathbf{x}'\mathbf{a}\mathbf{x}).$$

On expansion, the coefficient of x_i^2 is a_{ii} and the coefficient of $x_i x_j (i \neq j)$ is $a_{ij} + a_{ji}$. It is therefore convenient to take $a_{ij} = a_{ji}$, so that the coefficient of $x_i x_j$ is $2a_{ij}$ (or $2a_{ji}$); this imposes no essential restriction on the form itself, but enables us to choose $\mathbf{a}$ as a

† This is, indeed, true for any scalar.

symmetrical matrix. Hence *in the theory of the quadratic form* $(\mathbf{x}'\mathbf{a}\mathbf{x})$ *we may always assume that the matrix* $\mathbf{a}$ *is symmetrical.*

We may now write the equation of a quadric in the simple form

$$(\mathbf{x}'\mathbf{a}\mathbf{x}) = 0,$$

where $\mathbf{a} = \mathbf{a}'$.

If, on the other hand, the matrix $\mathbf{a}$ is skew-symmetrical, then $(\mathbf{x}'\mathbf{a}\mathbf{x}) \equiv 0$; for, by definition, the coefficients a_{ii} and $a_{ij} + a_{ji}$ are all zero.

10. Polar theory of the quadric. Let Q be the quadric whose equation is

$$Q \equiv (\mathbf{x}'\mathbf{a}\mathbf{x}) = 0.$$

The line joining the points $\mathbf{x}$, $\mathbf{y}$ meets the quadric in a point

$$\mathbf{z} \equiv \lambda\mathbf{x} + \mu\mathbf{y},$$

where

$$(\mathbf{z}'\mathbf{a}\mathbf{z}) = 0,$$

or

$$(\lambda\mathbf{x}' + \mu\mathbf{y}', \mathbf{a}, \lambda\mathbf{x} + \mu\mathbf{y}) = 0.$$

Hence

$$(\mathbf{x}'\mathbf{a}\mathbf{x})\lambda^2 + \{(\mathbf{x}'\mathbf{a}\mathbf{y}) + (\mathbf{y}'\mathbf{a}\mathbf{x})\}\lambda\mu + (\mathbf{y}'\mathbf{a}\mathbf{y})\mu^2 = 0,$$

or

$$(\mathbf{x}'\mathbf{a}\mathbf{x})\lambda^2 + 2(\mathbf{y}'\mathbf{a}\mathbf{x})\lambda\mu + (\mathbf{y}'\mathbf{a}\mathbf{y})\mu^2 = 0.$$

This is the matrix form of *Joachimstal's equation*, and the polar theory follows from it exactly as in the earlier part of the book. In particular, the tangent plane at a point $\mathbf{y}$ of Q, or the polar plane of a point $\mathbf{y}$ not on Q, is given by the equation

$$(\mathbf{y}'\mathbf{a}\mathbf{x}) = 0,$$

and two points $\mathbf{y}$, $\mathbf{z}$ are conjugate with respect to Q if

$$(\mathbf{y}'\mathbf{a}\mathbf{z}) = 0 \quad \text{or} \quad (\mathbf{z}'\mathbf{a}\mathbf{y}) = 0.$$

The plane vector $\mathbf{u}$ of the polar plane of $\mathbf{y}$ is found by identifying the equations $(\mathbf{u}'\mathbf{x}) = 0$, $(\mathbf{y}'\mathbf{a}\mathbf{x}) = 0$, so that

$$\mathbf{u}' = \mathbf{y}'\mathbf{a},$$

or

$$\mathbf{u} = \mathbf{a}\mathbf{y} \quad (\mathbf{a} = \mathbf{a}').$$

Hence *the polar plane of* $\mathbf{y}$ *has plane vector* $\mathbf{a}\mathbf{y}$.

Now suppose that $\mathbf{l}$ is a given plane. We find the point vector of its pole. In fact, if the pole is $\mathbf{y}$, then, by what we have just done,

$$\mathbf{l} = \mathbf{ay},$$

or, if Q is non-singular, $\qquad \mathbf{y} = \mathbf{a}^{-1}\mathbf{l}.$

It is customary to express this result in slightly different form by means of the standard relations $\mathbf{aA} = \mathbf{Aa} = |\mathbf{a}|\mathbf{I}$, where $\mathbf{A}$ is the adjoint of $\mathbf{a}$. We have

$$\mathbf{y} = \frac{\mathbf{A}}{|\mathbf{a}|}\mathbf{l},$$

so that, absorbing a constant since the coordinates are homogeneous, *the pole* $\mathbf{y}$ *of the plane* $\mathbf{l}$ *is given by the relation*

$$\mathbf{y} = \mathbf{Al}.$$

We can now find *the tangential equation of the quadric* Q. Taking $\mathbf{u}$ as the current vector of a plane touching Q, the pole of $\mathbf{u}$ is $\mathbf{Au}$. In order that the plane should touch the quadric, this point must also lie in its polar plane. Hence

$$(\mathbf{u}'\mathbf{Au}) = 0,$$

which is the required equation.

A similar argument enables us to find *the reciprocal of the quadric* $Q \equiv (\mathbf{x}'\mathbf{ax}) = 0$ *with respect to the quadric* $R \equiv (\mathbf{x}'\mathbf{bx}) = 0$.

If $\mathbf{y}$ is a typical point of Q, so that

$$(\mathbf{y}'\mathbf{ay}) = 0,$$

its polar plane with respect to R is $\mathbf{u}$, where

$$\mathbf{u} = \mathbf{by},$$

or $\qquad \mathbf{y} = \mathbf{b}^{-1}\mathbf{u}.$

For convenience, replace $\mathbf{b}^{-1}$ by $\dfrac{\mathbf{B}}{|\mathbf{b}|}$, where $\mathbf{B}$ is, as usual, the adjoint of $\mathbf{b}$. Then, by substitution in the equation $(\mathbf{y}'\mathbf{ay}) = 0$,

$$(\mathbf{u}'\mathbf{BaBu}) = 0 \qquad (\mathbf{B} = \mathbf{B}').$$

Thus *the reciprocal of the quadric of matrix* $\mathbf{a}$ *with respect to the quadric of matrix* $\mathbf{b}$ *is the quadric envelope of matrix* $\mathbf{BaB}$.

ILLUSTRATION 2. *Prove that the joins of a point* $\mathbf{y}$ *to the points of the section of the quadric* $(\mathbf{x}'\mathbf{ax}) = 0$ *by the plane* $(\mathbf{l}'\mathbf{x}) = 0$ *meet the quadric again in points lying in the plane* $(\mathbf{m}'\mathbf{x}) = 0$, *where*

$$\mathbf{m} \equiv 2(\mathbf{l}'\mathbf{y})\,\mathbf{ay} - (\mathbf{y}'\mathbf{ay})\,\mathbf{l}.$$

If the plane $\mathbf{l}$ *is fixed and* $\mathbf{y}$ *varies in a fixed plane* $(\mathbf{p}'\mathbf{x}) = 0$, *prove that the locus of the pole of the plane* $\mathbf{m}$ *with respect to the quadric is the quadric*

$$(\mathbf{p}'\mathbf{Al})^2\,(\mathbf{x}'\mathbf{ax}) - |\,\mathbf{a}\,|\,(\mathbf{l}'\mathbf{Al})\,(\mathbf{p}'\mathbf{x})^2 = 0,$$

where $\mathbf{A}$ *is the adjoint of the matrix* $\mathbf{a}$. [P.]

(i) Let $\mathbf{z}$ be a point on the section of the quadric by the plane, so that

$$(\mathbf{z}'\mathbf{az}) = 0, \quad (\mathbf{l}'\mathbf{z}) = 0.$$

The line joining $\mathbf{y}$, $\mathbf{z}$ meets the quadric in two points of the form $\lambda\mathbf{y} + \mu\mathbf{z}$, where, according to Joachimstal's equation,

$$(\mathbf{y}'\mathbf{ay})\,\lambda^2 + 2(\mathbf{y}'\mathbf{az})\,\lambda\mu + (\mathbf{z}'\mathbf{az})\,\mu^2 = 0.$$

Since $(\mathbf{z}'\mathbf{az}) = 0$, one root is $\lambda = 0$; for the point, say $\boldsymbol{\xi}$, given by the other, we have

$$(\mathbf{y}'\mathbf{ay})\,\lambda + 2(\mathbf{y}'\mathbf{az})\,\mu = 0,$$

so that

$$\boldsymbol{\xi} = 2(\mathbf{y}'\mathbf{az})\,\mathbf{y} - (\mathbf{y}'\mathbf{ay})\,\mathbf{z}.$$

But $(\mathbf{l}'\mathbf{z}) = 0$. Hence $\quad (\mathbf{l}'\boldsymbol{\xi}) = 2(\mathbf{y}'\mathbf{az})\,(\mathbf{l}'\mathbf{y})$.

Also

$$(\mathbf{y}'\mathbf{a}\boldsymbol{\xi}) = 2(\mathbf{y}'\mathbf{az})\,(\mathbf{y}'\mathbf{ay}) - (\mathbf{y}'\mathbf{ay})\,(\mathbf{y}'\mathbf{az})$$
$$= (\mathbf{y}'\mathbf{az})\,(\mathbf{y}'\mathbf{ay}).$$

Eliminating $(\mathbf{y}'\mathbf{az})$ between the last two equations, we have the relation

$$2(\mathbf{l}'\mathbf{y})\,(\mathbf{y}'\mathbf{a}\boldsymbol{\xi}) - (\mathbf{y}'\mathbf{ay})\,(\mathbf{l}'\boldsymbol{\xi}) = 0,$$

and so the point $\boldsymbol{\xi}$ lies in the plane $(\mathbf{m}'\mathbf{x}) = 0$, where

$$\mathbf{m}' = 2(\mathbf{l}'\mathbf{y})\,\mathbf{y}'\mathbf{a} - (\mathbf{y}'\mathbf{ay})\,\mathbf{l}'.$$

Transposing, and remembering that expressions in brackets are scalars, we obtain the equation

$$\mathbf{m} = 2(\mathbf{l}'\mathbf{y})\,\mathbf{ay} - (\mathbf{y}'\mathbf{ay})\,\mathbf{l} \qquad (\mathbf{a}' = \mathbf{a}).$$

(ii) We have $$(\mathbf{p}'\mathbf{y}) = 0.$$

The pole of the plane $\mathbf{m}$ with respect to $(\mathbf{x}'\mathbf{ax}) = 0$ may be taken as $\mathbf{x}$, where

$$\mathbf{x} = \mathbf{a}^{-1}\mathbf{m} = 2(\mathbf{l}'\mathbf{y})\,\mathbf{y} - (\mathbf{y}'\mathbf{ay})\,\mathbf{a}^{-1}\mathbf{l}.$$

Hence $$(\mathbf{p}'\mathbf{x}) = -(\mathbf{y}'\mathbf{ay})\,(\mathbf{p}'\mathbf{a}^{-1}\mathbf{l}).$$

Also $$\mathbf{ax} = 2(\mathbf{l}'\mathbf{y})\,\mathbf{ay} - (\mathbf{y}'\mathbf{ay})\,\mathbf{l},$$

so that $$\mathbf{x}'\mathbf{a} = 2(\mathbf{l}'\mathbf{y})\,\mathbf{y}'\mathbf{a} - (\mathbf{y}'\mathbf{ay})\,\mathbf{l}',$$

and*
$$\begin{aligned}(\mathbf{x}'\mathbf{ax}) &= \{2(\mathbf{l}'\mathbf{y})\,\mathbf{y}'\mathbf{a},\ 2(\mathbf{l}'\mathbf{y})\,\mathbf{y} - (\mathbf{y}'\mathbf{ay})\,\mathbf{a}^{-1}\mathbf{l}\}\\ &\quad - \{(\mathbf{y}'\mathbf{ay})\,\mathbf{l}',\ 2(\mathbf{l}'\mathbf{y})\,\mathbf{y} - (\mathbf{y}'\mathbf{ay})\,\mathbf{a}^{-1}\mathbf{l}\}\\ &= 4(\mathbf{l}'\mathbf{y})^2(\mathbf{y}'\mathbf{ay}) - 2(\mathbf{l}'\mathbf{y})(\mathbf{y}'\mathbf{ay})(\mathbf{y}'\mathbf{l})\\ &\quad - 2(\mathbf{y}'\mathbf{ay})(\mathbf{l}'\mathbf{y})^2 + (\mathbf{y}'\mathbf{ay})^2(\mathbf{l}'\mathbf{a}^{-1}\mathbf{l})\\ &= (\mathbf{y}'\mathbf{ay})^2(\mathbf{l}'\mathbf{a}^{-1}\mathbf{l}),\end{aligned}$$

since $$(\mathbf{l}'\mathbf{y}) = (\mathbf{y}'\mathbf{l}).$$

The two relations
$$(\mathbf{p}'\mathbf{x}) = -(\mathbf{y}'\mathbf{ay})\,(\mathbf{p}'\mathbf{a}^{-1}\mathbf{l}),$$
$$(\mathbf{x}'\mathbf{ax}) = (\mathbf{y}'\mathbf{ay})^2\,(\mathbf{l}'\mathbf{a}^{-1}\mathbf{l})$$

give $$(\mathbf{p}'\mathbf{a}^{-1}\mathbf{l})^2\,(\mathbf{x}'\mathbf{ax}) = (\mathbf{l}'\mathbf{a}^{-1}\mathbf{l})\,(\mathbf{p}'\mathbf{x})^2,$$

and, replacing $\mathbf{a}^{-1}$ by $\dfrac{\mathbf{A}}{|\mathbf{a}|}$, we have the required equation

$$(\mathbf{p}'\mathbf{Al})^2\,(\mathbf{x}'\mathbf{ax}) = |\mathbf{a}|\,(\mathbf{l}'\mathbf{Al})\,(\mathbf{p}'\mathbf{x})^2.$$

ILLUSTRATION 3. *The equations of two quadric surfaces are* $(\mathbf{x}'\mathbf{ax}) = 0$, $(\mathbf{x}'\mathbf{bx}) = 0$, *where both* $\mathbf{a}$ *and* $\mathbf{b}$ *are symmetric and non-singular matrices. Show that either quadric is its own reciprocal with respect to the other when, and only when, the square of* $\mathbf{ba}^{-1}$ *is a scalar multiple of the unit matrix.*

A skew quadrilateral q is formed by taking a pair of lines from each regulus on a quadric Q. Show that there is one, and only one, other quadric that contains q and is at the same time its own reciprocal with respect to Q. [M.T. II.]

(i) We first find the reciprocal of $(\mathbf{x}'\mathbf{ax}) = 0$ with respect to $(\mathbf{x}'\mathbf{bx}) = 0$. Let $\mathbf{u}$ be a tangent plane of $(\mathbf{x}'\mathbf{ax}) = 0$, so that

$$(\mathbf{u}'\mathbf{a}^{-1}\mathbf{u}) = 0.$$

* The commas and { } brackets are used for clarity in a complicated expression.

If its pole with respect to $(\mathbf{x}'\mathbf{bx}) = 0$ is $\mathbf{y}$, then

$$\mathbf{u} = \mathbf{by}.$$

Hence $$(\mathbf{y}'\mathbf{b}'\mathbf{a}^{-1}\mathbf{by}) = 0,$$

or $$(\mathbf{y}'\mathbf{ba}^{-1}\mathbf{by}) = 0 \qquad (\mathbf{b}' = \mathbf{b}).$$

The reciprocal of $(\mathbf{x}'\mathbf{ax})$ with respect to $(\mathbf{x}'\mathbf{bx})$, being the locus of $\mathbf{y}$, is therefore given by the equation

$$(\mathbf{x}'\mathbf{ba}^{-1}\mathbf{bx}) = 0.$$

This is $(\mathbf{x}'\mathbf{ax}) = 0$ itself if, and only if, a scalar λ exists such that

$$\mathbf{ba}^{-1}\mathbf{b} = \lambda\mathbf{a},$$

each matrix being symmetric. Hence

$$\mathbf{ba}^{-1}\mathbf{ba}^{-1} = \lambda\mathbf{I},$$

as required.

Note that, when this condition is satisfied, then

$$\mathbf{ab}^{-1}\mathbf{ab}^{-1} = \frac{1}{\lambda}\mathbf{I},$$

so that the defining property holds for *either* quadric.

(ii) We may take Q in the form

$$Q \equiv 2a_{03}x_0x_3 + 2a_{12}x_1x_2 = 0,$$

the sides of q being

$$x_0 = x_1 = 0, \quad x_0 = x_2 = 0, \quad x_3 = x_1 = 0, \quad x_3 = x_2 = 0.$$

If the second quadric, say R, exists, its equation is

$$R \equiv 2b_{03}x_0x_3 + 2b_{12}x_1x_2 = 0.$$

Then $$|\mathbf{a}| = a_{03}^2a_{12}^2, \quad |\mathbf{b}| = b_{03}^2b_{12}^2,$$

and

$$\mathbf{a}^{-1} = \begin{pmatrix} 0 & 0 & 0 & \dfrac{1}{a_{03}} \\ 0 & 0 & \dfrac{1}{a_{12}} & 0 \\ 0 & \dfrac{1}{a_{12}} & 0 & 0 \\ \dfrac{1}{a_{03}} & 0 & 0 & 0 \end{pmatrix},$$

so that

$$\mathbf{ba}^{-1} = \begin{pmatrix} \frac{b_{03}}{a_{03}} & 0 & 0 & 0 \\ 0 & \frac{b_{12}}{a_{12}} & 0 & 0 \\ 0 & 0 & \frac{b_{12}}{a_{12}} & 0 \\ 0 & 0 & 0 & \frac{b_{03}}{a_{03}} \end{pmatrix}.$$

Hence $\mathbf{ba}^{-1}\mathbf{ba}^{-1} \equiv \mathbf{I}$ (absorbing the scalar factor) if

$$b_{03}^2 = a_{03}^2, \quad b_{12}^2 = a_{12}^2,$$

so that $$b_{03} = \pm a_{03}, \quad b_{12} = \pm a_{12}.$$

The only quadric distinct from Q is therefore

$$2a_{03}x_0x_3 - 2a_{12}x_1x_2 = 0.$$

11. Line-coordinates. Let $\mathbf{x}$, $\mathbf{y}$ be two points on a given line p. The 4×4 matrix

$$\mathbf{p} \equiv \mathbf{xy}' - \mathbf{yx}'$$

is defined as the *coordinate matrix* of the line.

If $\mathbf{x}$, $\mathbf{y}$ are replaced by two other points $\boldsymbol{\xi} \equiv \lambda\mathbf{x} + \mu\mathbf{y}$, $\boldsymbol{\eta} \equiv \nu\mathbf{x} + \rho\mathbf{y}$ of the line, we have

$$\begin{aligned} \boldsymbol{\xi\eta}' - \boldsymbol{\eta\xi}' &\equiv (\lambda\mathbf{x} + \mu\mathbf{y})(\nu\mathbf{x}' + \rho\mathbf{y}') - (\nu\mathbf{x} + \rho\mathbf{y})(\lambda\mathbf{x}' + \mu\mathbf{y}') \\ &= (\lambda\rho - \mu\nu)(\mathbf{xy}' - \mathbf{yx}') \\ &= (\lambda\rho - \mu\nu)\,\mathbf{p}, \end{aligned}$$

so that $\mathbf{p}$ is determined by the line, to within a scalar factor.

Next, let $\mathbf{u}$, $\mathbf{v}$ be two planes through p. The 4×4 matrix

$$\boldsymbol{\pi} \equiv \mathbf{uv}' - \mathbf{vu}'$$

is defined as the *dual coordinate matrix* of the line. As for $\mathbf{p}$, we can prove that $\boldsymbol{\pi}$ is determined by the line, to within a scalar factor.

The two matrices are skew-symmetrical, since

$$\mathbf{p}' = \mathbf{yx}' - \mathbf{xy}' = -\mathbf{p},$$

$$\boldsymbol{\pi}' = \mathbf{vu}' - \mathbf{uv}' = -\boldsymbol{\pi}.$$

The matrices $\mathbf{p}$, $\boldsymbol{\pi}$ are not independent; in fact,

$$\begin{aligned}\mathbf{p}\boldsymbol{\pi} &= \mathbf{x}\mathbf{y}'\mathbf{u}\mathbf{v}' - \mathbf{x}\mathbf{y}'\mathbf{v}\mathbf{u}' - \mathbf{y}\mathbf{x}'\mathbf{u}\mathbf{v}' + \mathbf{y}\mathbf{x}'\mathbf{v}\mathbf{u}' \\ &= (\mathbf{y}'\mathbf{u})\,\mathbf{x}\mathbf{v}' - (\mathbf{y}'\mathbf{v})\,\mathbf{x}\mathbf{u}' - (\mathbf{x}'\mathbf{u})\,\mathbf{y}\mathbf{v}' + (\mathbf{x}'\mathbf{v})\,\mathbf{y}\mathbf{u}',\end{aligned}$$

on taking out scalar factors. But each of the points $\mathbf{x}$, $\mathbf{y}$ lies on p and therefore in each of the planes $\mathbf{u}$, $\mathbf{v}$. Hence

$$(\mathbf{y}'\mathbf{u}) = (\mathbf{y}'\mathbf{v}) = (\mathbf{x}'\mathbf{u}) = (\mathbf{x}'\mathbf{v}) = 0,$$

and so *the matrices* $\mathbf{p}$, $\boldsymbol{\pi}$ *are connected by the fundamental relation*

$$\mathbf{p}\boldsymbol{\pi} = 0.$$

Two distinct sets of results follow from this equation. Considering the component in the ith row and the jth column, we have the relation

$$[\mathbf{p}\boldsymbol{\pi}]_{ij} = \sum_{\lambda=0}^{3} p_{i\lambda}\pi_{\lambda j}.$$

(i) If $i \neq j$, then this relation is

$$p_{ik}\pi_{kj} + p_{il}\pi_{lj} = 0,$$

where k, l are different from i, j. Hence

$$p_{ik}\pi_{kj} = p_{il}\pi_{jl},$$

or

$$\frac{p_{ik}}{\pi_{jl}} = \frac{p_{il}}{\pi_{kj}}.$$

Taking successive particular values for i, j, k, l, we reach the standard relations

$$\frac{p_{01}}{\pi_{23}} = \frac{p_{02}}{\pi_{31}} = \frac{p_{03}}{\pi_{12}} = \frac{p_{23}}{\pi_{01}} = \frac{p_{31}}{\pi_{02}} = \frac{p_{12}}{\pi_{03}}.$$

(ii) If $i = j$, the relation is

$$p_{ik}\pi_{ki} + p_{il}\pi_{li} + p_{im}\pi_{mi} = 0,$$

where k, l, m are different from i.

But, by (i), we have

$$\frac{\pi_{ki}}{p_{lm}} = \frac{\pi_{li}}{p_{mk}} = \frac{\pi_{mi}}{p_{kl}}.$$

Hence

$$p_{ik}p_{lm} + p_{il}p_{mk} + p_{im}p_{kl} = 0.$$

The different sets of (distinct) values of i, k, l, m all yield the same fundamental quadratic relation

$$p_{01}p_{23}+p_{02}p_{31}+p_{03}p_{12}=0.$$

Similarly $$\pi_{01}\pi_{23}+\pi_{02}\pi_{31}+\pi_{03}\pi_{12}=0.$$

These matrices are clearly connected with the line-coordinates l, m, n, l', m', n' discussed earlier. In fact, we may compare results by means of the 'dictionary'

$$l=p_{10},\quad l'=p_{23},$$
$$m=p_{20},\quad m'=p_{31},$$
$$n=p_{30},\quad n'=p_{12},$$

if we equate the coefficients x, y, z, t of the former notation to the present x_1, x_2, x_3, x_0.

The point where a line $\mathbf{p}$ meets a plane $\mathbf{u}$ is easily determined. Consider the matrix

$$\boldsymbol{\xi}\equiv\mathbf{pu}.$$

By definition of $\mathbf{p}$, we have

$$\boldsymbol{\xi}=\mathbf{xy'u}-\mathbf{yx'u}$$
$$=(\mathbf{y'u})\,\mathbf{x}-(\mathbf{x'u})\,\mathbf{y},$$

which is the algebraic sum of a scalar multiple of $\mathbf{x}$ and a scalar multiple of $\mathbf{y}$. Hence $\boldsymbol{\xi}$ lies on the line $\mathbf{p}$ joining $\mathbf{x}$, $\mathbf{y}$.

Moreover, since $\mathbf{p}$ is skew-symmetrical, the bilinear form $(\mathbf{u'pu})$ vanishes, so that

$$(\mathbf{u'}\boldsymbol{\xi})=0.$$

Hence $\boldsymbol{\xi}$ lies in the plane $\mathbf{u}$.

It follows that $\boldsymbol{\xi}$, *where* $\boldsymbol{\xi}\equiv\mathbf{pu}$,

is the point of intersection of the line $\mathbf{p}$ *with the plane* $\mathbf{u}$.

Dually, we may show that $\boldsymbol{\alpha}$, *where*

$$\boldsymbol{\alpha}\equiv\boldsymbol{\pi}\mathbf{x},$$

is the plane joining the line $\mathbf{p}$ *to the point* $\mathbf{x}$.

Note also that, from the equation

$$\mathbf{pu}\equiv\boldsymbol{\xi}=(\mathbf{y'u})\,\mathbf{x}-(\mathbf{x'u})\,\mathbf{y},$$

the line $\mathbf{p}$ *lies in the plane* $\mathbf{u}$ *if* $\mathbf{pu}=\mathbf{0}$.

Similarly, *the line* $\mathbf{p}$ *passes through the point* $\mathbf{x}$ *if* $\boldsymbol{\pi}\mathbf{x}=\mathbf{0}$.

As a corollary of the last remark, *the four column vectors, conveniently denoted by* $\mathbf{p}_0, \mathbf{p}_1, \mathbf{p}_2, \mathbf{p}_3$, *whose elements are, in succession, the four columns of* $\mathbf{p}$, *denote four points lying on* $\mathbf{p}$ (except, of course, for any $\mathbf{p}_i$ which, exceptionally, is zero).

In fact, $$\mathbf{p}_i \equiv \{p_{0i}, p_{1i}, p_{2i}, p_{3i}\};$$
and the relation $$\mathbf{pu} = \mathbf{0}$$
is the same as $$\mathbf{u'p'} = \mathbf{0}$$
or $$\mathbf{u'p} = \mathbf{0} \qquad (\mathbf{p'} = -\mathbf{p}).$$
Taking components, we have
$$\sum_{\lambda} u'_{\lambda} p_{\lambda i} = 0 \quad (i = 0, 1, 2, 3)$$
or $$(\mathbf{u'p}_i) = 0.$$
This means that, if $\mathbf{u}$ is *any* plane through $\mathbf{p}$, the point $\mathbf{p}_i$ lies in it. Hence $\mathbf{p}_i$ is a point of $\mathbf{p}$.

In the same way, *the four column vectors* $\boldsymbol{\pi}_0, \boldsymbol{\pi}_1, \boldsymbol{\pi}_2, \boldsymbol{\pi}_3$ *of* $\boldsymbol{\pi}$ *denote four planes passing through* $\mathbf{p}$ (except for any $\boldsymbol{\pi}_i$ which, exceptionally, is zero).

Since there are only two linearly independent vectors for the points of a line, only two of $\mathbf{p}_0, \mathbf{p}_1, \mathbf{p}_2, \mathbf{p}_3$ are independent. Hence the determinant of their coefficients vanishes; that is,
$$|\mathbf{p}| = 0.$$
Similarly $$|\boldsymbol{\pi}| = 0.$$

We can, in fact, prove easily that
$$|\mathbf{p}| = (p_{01}p_{23} + p_{02}p_{31} + p_{03}p_{12})^2,$$
remembering that $\mathbf{p}$ is skew-symmetrical. The result $|\mathbf{p}| = 0$ is thus the same as the quadratic condition $p_{01}p_{23} + p_{02}p_{31} + p_{03}p_{12} = 0$ obtained before.

In order to proceed further, we require an algebraic lemma:

If $\mathbf{p}_i, \mathbf{p}_j$ *are two different column vectors from the matrix* $\mathbf{p}$, *then*
$$\mathbf{p}_i\mathbf{p}'_j - \mathbf{p}_j\mathbf{p}'_i = p_{ij}\mathbf{p}.$$

The element in the ath row and bth column on the left-hand side is

$$\begin{aligned} & p_{ai}p_{bj} - p_{aj}p_{bi} \\ &= -p_{ai}p_{jb} - p_{aj}p_{bi} \qquad (p_{bj} = -p_{jb}) \\ &= p_{ab}p_{ij} \end{aligned}$$

by the fundamental quadratic relation when a, b, i, j are distinct, and trivially otherwise. But this is p_{ij} times the corresponding element of $\mathbf{p}$, which proves the lemma.

Note that the relation $\mathbf{p}_i\mathbf{p}_j' - \mathbf{p}_j\mathbf{p}_i' = k\mathbf{p}$ follows from the definition of $\mathbf{p}$ since $\mathbf{p}_i, \mathbf{p}_j$ are points of the line. The lemma evaluates the constant k.

We can now prove that, *if* $\mathbf{q}$ *is a skew-symmetric matrix whose determinant vanishes, then there exists a unique line of which* $\mathbf{q}$ *is the coordinate matrix.*

To give anything to talk about, $\mathbf{q}$ must have at least one non-zero element, say q_{ij}. Then q_{ji} is also non-zero. Let $\mathbf{q}_i$, $\mathbf{q}_j$ denote the ith and jth column vectors from $\mathbf{q}$; by selection, they are not identically zero, and so they are the coordinates of two points, necessarily distinct.

Consider the line joining the points $\mathbf{q}_i$, $\mathbf{q}_j$. It certainly exists, and its coordinate matrix is $\mathbf{p}$, where

$$\begin{aligned} \mathbf{p} &= \mathbf{q}_i\mathbf{q}_j' - \mathbf{q}_j\mathbf{q}_i' \\ &= q_{ij}\mathbf{q}, \end{aligned}$$

by the lemma. Hence the coordinate matrix of this line is a (non-zero) scalar multiple of $\mathbf{q}$.

Moreover, the line is unique; for any line whose coordinate matrix is $\mathbf{q}$ must contain the two distinct points $\mathbf{q}_i$, $\mathbf{q}_j$; and there is only one such line.

We pass to *two given straight lines*, and require the condition that they should intersect. Let the two lines have coordinate matrices $\mathbf{p}$, $\mathbf{p}^*$ and dual coordinate matrices π, π^*.

Suppose that the two lines meet in the point $\mathbf{x}$ and lie in the plane $\mathbf{u}$. Take further points $\mathbf{y}$, $\mathbf{y}^*$ on them and planes $\mathbf{v}$, $\mathbf{v}^*$ through them. Then

$$\mathbf{p} = \mathbf{x}\mathbf{y}' - \mathbf{y}\mathbf{x}', \quad \mathbf{p}^* = \mathbf{x}\mathbf{y}^{*\prime} - \mathbf{y}^*\mathbf{x}',$$

$$\pi = \mathbf{u}\mathbf{v}' - \mathbf{v}\mathbf{u}', \quad \pi^* = \mathbf{u}\mathbf{v}^{*\prime} - \mathbf{v}^*\mathbf{u}'.$$

Remembering that $\mathbf{x}$ lies in $\mathbf{u}, \mathbf{v}, \mathbf{v}^*$; $\mathbf{y}$ in $\mathbf{u}, \mathbf{v}$; and $\mathbf{y}^*$ in $\mathbf{u}, \mathbf{v}^*$, we have the relations

$$\mathbf{p}\boldsymbol{\pi}^* = -(\mathbf{y}'\mathbf{v}^*)\,\mathbf{x}\mathbf{u}',$$

$$\mathbf{p}^*\boldsymbol{\pi} = -(\mathbf{y}^{*\prime}\mathbf{v})\,\mathbf{x}\mathbf{u}'.$$

[It is tempting at this point to note that $\mathbf{p}\boldsymbol{\pi}^*$ and $\mathbf{p}^*\boldsymbol{\pi}$ are both scalar multiples of $\mathbf{x}\mathbf{u}'$, and so to seek a relation which, for symmetry, we should like to be $\mathbf{p}\boldsymbol{\pi}^* + \mathbf{p}^*\boldsymbol{\pi} = 0$. But this cannot be, since $\mathbf{p}, \mathbf{p}^*, \boldsymbol{\pi}, \boldsymbol{\pi}^*$ are all, in their very definitions, arbitrary to the extent of a scalar factor. We can, indeed, 'normalise' them, by taking each of the six equal ratios such as p_{01}/π_{23} to be unity, and the relation follows. This is just the '$lmnl'm'n'$' method used before. But the matrix treatment becomes concealed in doing so.]

From the two matrix equations we can derive *four equivalent expressions for the condition that the lines should intersect:*

Since
$$(\mathbf{y}^{*\prime}\mathbf{v})\,\mathbf{p}\boldsymbol{\pi}^* = (\mathbf{y}'\mathbf{v}^*)\,\mathbf{p}^*\boldsymbol{\pi},$$
we have
$$(\mathbf{y}^{*\prime}\mathbf{v})\,\mathbf{p}\boldsymbol{\pi}^*\mathbf{p} = (\mathbf{y}'\mathbf{v}^*)\,\mathbf{p}^*\boldsymbol{\pi}\mathbf{p} = \mathbf{0},$$

since $\boldsymbol{\pi}\mathbf{p} = \mathbf{0}$. But $(\mathbf{y}^{*\prime}\mathbf{v}) \neq 0$. Hence $\mathbf{p}\boldsymbol{\pi}^*\mathbf{p} = \mathbf{0}$. By similar reasoning, we obtain the four forms

$$\begin{cases} \mathbf{p}\boldsymbol{\pi}^*\mathbf{p} = \mathbf{0}, & \boldsymbol{\pi}\mathbf{p}^*\boldsymbol{\pi} = \mathbf{0}, \\ \mathbf{p}^*\boldsymbol{\pi}\mathbf{p}^* = \mathbf{0}, & \boldsymbol{\pi}^*\mathbf{p}\boldsymbol{\pi}^* = \mathbf{0}. \end{cases}$$

Suppose, conversely, that $\mathbf{p}\boldsymbol{\pi}^*\mathbf{p} = \mathbf{0}$. Then $\mathbf{p}\,.\,\boldsymbol{\pi}^*\mathbf{p} = \mathbf{0}$, so that

$$\mathbf{p}\,.\,\boldsymbol{\pi}^*\mathbf{p}_i = \mathbf{0} \quad (i = 0, 1, 2, 3).$$

But $\boldsymbol{\pi}^*\mathbf{p}_i$ is the plane containing the line $\mathbf{p}^*$ and the point $\mathbf{p}_i$; the equation just given shows that the line $\mathbf{p}$ lies in this plane. Hence the two lines intersect.

We now show that the equation $\mathbf{p}\boldsymbol{\pi}^*\mathbf{p} = \mathbf{0}$ is equivalent to the earlier condition (p. 67)

$$l_1 l_2' + l_2 l_1' + m_1 m_2' + m_2 m_1' + n_1 n_2' + n_2 n_1' = 0$$

for the lines to intersect.

In order to do this, we prove a slightly more general result, which will be used again shortly. Let $\mathbf{a}$ be any skew-symmetric matrix (not necessarily of zero determinant) and consider the product

$$\begin{aligned} \mathbf{pap} &= [\mathbf{xy}' - \mathbf{yx}']\,\mathbf{a}[\mathbf{xy}' - \mathbf{yx}'] \\ &= \mathbf{xy}'\mathbf{axy}' - \mathbf{xy}'\mathbf{ayx}' - \mathbf{yx}'\mathbf{axy}' + \mathbf{yx}'\mathbf{ayx}' \\ &= (\mathbf{y}'\mathbf{ax})\,\mathbf{xy}' - (\mathbf{y}'\mathbf{ay})\,\mathbf{xx}' - (\mathbf{x}'\mathbf{ax})\,\mathbf{yy}' + (\mathbf{x}'\mathbf{ay})\,\mathbf{yx}'. \end{aligned}$$

Since $\mathbf{a}$ is skew-symmetric,

$$(\mathbf{x}'\mathbf{a}\mathbf{x}) = (\mathbf{y}'\mathbf{a}\mathbf{y}) = 0,$$

and

$$(\mathbf{y}'\mathbf{a}\mathbf{x}) = -(\mathbf{x}'\mathbf{a}\mathbf{y}),$$

and so

$$\mathbf{pap} = (\mathbf{y}'\mathbf{a}\mathbf{x})\,[\mathbf{x}\mathbf{y}' - \mathbf{y}\mathbf{x}'] = (\mathbf{y}'\mathbf{a}\mathbf{x})\,\mathbf{p} = -(\mathbf{x}'\mathbf{a}\mathbf{y})\,\mathbf{p}.$$

Now

$$(\mathbf{x}'\mathbf{a}\mathbf{y}) = \sum_\lambda \sum_\mu x_\lambda a_{\lambda\mu} y_\mu,$$

$$(\mathbf{y}'\mathbf{a}\mathbf{x}) = \sum_\lambda \sum_\mu y_\lambda a_{\lambda\mu} x_\mu.$$

Subtracting,

$$2(\mathbf{x}'\mathbf{a}\mathbf{y}) = \sum_\lambda \sum_\mu a_{\lambda\mu}(x_\lambda y_\mu - y_\lambda x_\mu) = \sum_\lambda \sum_\mu a_{\lambda\mu} p_{\lambda\mu}$$
$$= 2(a_{01}p_{01} + a_{02}p_{02} + a_{03}p_{03} + a_{23}p_{23} + a_{31}p_{31} + a_{12}p_{12}),$$

since each term occurs twice in the summation on the right-hand side. Hence

$$\mathbf{pap} = -(a_{01}p_{01} + a_{02}p_{02} + a_{03}p_{03} + a_{23}p_{23} + a_{31}p_{31} + a_{12}p_{12})\,\mathbf{p}.$$

Returning to our immediate problem, consider the case when $\mathbf{a}$ is $\boldsymbol{\pi}^*$. Then the relation $\mathbf{p}\boldsymbol{\pi}^*\mathbf{p} = \mathbf{0}$ is equivalent to

$$\varpi \mathbf{p} = \mathbf{0},$$

where

$$\varpi \equiv \pi^*_{01}p_{01} + \pi^*_{02}p_{02} + \pi^*_{03}p_{03} + \pi^*_{23}p_{23} + \pi^*_{31}p_{31} + \pi^*_{12}p_{12}.$$

Since $\mathbf{p}$ is not zero, we have the condition

$$\varpi = 0$$

for two intersecting lines.

In terms of coordinate matrices $\mathbf{p}$, $\mathbf{p}^*$, the condition is

$$p_{01}p^*_{23} + p_{23}p^*_{01} + p_{02}p^*_{31} + p_{31}p^*_{02} + p_{03}p^*_{12} + p_{12}p^*_{03} = 0.$$

This is as far as we propose to go by these methods. We hope the reader has been tempted to procure a more advanced account of the work. To conclude, we give a brief note on the linear complex.

By definition, a linear complex consists of those lines whose coordinates are subject to a linear relation, say

$$\sum_\lambda \sum_\mu a_{\lambda\mu} p_{\lambda\mu} = 0,$$

where the matrix $\mathbf{a}$ may be taken as skew-symmetric without loss of generality.

By what we have just proved, this is the same as the matrix equation

$$\mathbf{pap} = \mathbf{0}.$$

[If, in addition, $|\mathbf{a}| = 0$, then the line $\mathbf{p}$ always meets the line whose dual coordinate matrix is $\mathbf{a}$. The complex is then said to be *special*.]

The above derivation of ϖ included the result that

$$\sum_{\lambda}\sum_{\mu} a_{\lambda\mu} p_{\lambda\mu} = 2(\mathbf{x}'\mathbf{ay});$$

the equation of the complex therefore shows that, if $\mathbf{x}$, $\mathbf{y}$ are two points on any line belonging to it, then

$$(\mathbf{x}'\mathbf{ay}) = 0,$$

or, equivalently,
$$(\mathbf{y}'\mathbf{ax}) = 0.$$

In particular, if we regard $\mathbf{y}$ as fixed, then *those lines of the complex which pass through a point* $\mathbf{y}$ *lie in the plane*

$$(\mathbf{y}'\mathbf{ax}) = 0.$$

This plane is called the *null plane of* $\mathbf{y}$ *with respect to the complex*. Its coordinate vector is $\mathbf{ay}$.

Dually, *those lines of the complex which lie in a plane pass through a point, called the null point of that plane.*

The relation $(\mathbf{y}'\mathbf{ay}) = 0$ shows that each point lies in its null plane.

The equation
$$(\mathbf{y}'\mathbf{ax}) = 0$$

is said to define a *null system*.

ILLUSTRATION 4. *To prove*† *that the equation of the quadric generated by the lines common to three non-singular linear complexes, of matrices* $\mathbf{a}$, $\mathbf{b}$, $\mathbf{c}$, *may be expressed in the form* $(\mathbf{x}'\mathbf{qx}) = 0$, *where*

$$\mathbf{q} \equiv \mathbf{cb}^{-1}\mathbf{a} - \mathbf{ab}^{-1}\mathbf{c}.$$

Let l be any line common to the three linear complexes; choose an arbitrary point $\boldsymbol{\alpha}$ of the line.

† See a paper by H. W. Turnbull, *Proceedings of the Cambridge Philosophical Society*, 31 (1935), p. 174.

The null plane of $\boldsymbol{\alpha}$ with respect to the complex $\mathbf{a}$ is $\mathbf{a}\boldsymbol{\alpha}$, and this plane contains the line l of $\mathbf{a}$.

The null point of $\mathbf{a}\boldsymbol{\alpha}$ with respect to the complex $\mathbf{b}$ is

$$\mathbf{b}^{-1}.\mathbf{a}\boldsymbol{\alpha} \equiv \mathbf{b}^{-1}\mathbf{a}\boldsymbol{\alpha},$$

and this point lies on the line l of $\mathbf{b}$.

The null plane of $\mathbf{b}^{-1}\mathbf{a}\boldsymbol{\alpha}$ with respect to the complex $\mathbf{c}$ is $\mathbf{c}\mathbf{b}^{-1}\mathbf{a}\boldsymbol{\alpha}$, and this plane contains the line l of $\mathbf{c}$.

In the same way, by reversing the order in which the complexes are selected, we prove that the plane $\mathbf{a}\mathbf{b}^{-1}\mathbf{c}\boldsymbol{\alpha}$ also contains l.

Hence the line l lies in each of the planes

$$(\mathbf{x}'\mathbf{c}\mathbf{b}^{-1}\mathbf{a}\boldsymbol{\alpha}) = 0,$$
$$(\mathbf{x}'\mathbf{a}\mathbf{b}^{-1}\mathbf{c}\boldsymbol{\alpha}) = 0.$$

But $\boldsymbol{\alpha}$ was defined as a point of l, and so

$$(\boldsymbol{\alpha}'\mathbf{c}\mathbf{b}^{-1}\mathbf{a}\boldsymbol{\alpha}) = 0,$$
$$(\boldsymbol{\alpha}'\mathbf{a}\mathbf{b}^{-1}\mathbf{c}\boldsymbol{\alpha}) = 0.$$

Hence $$(\boldsymbol{\alpha}'\mathbf{q}\boldsymbol{\alpha}) = 0,$$

where $$\mathbf{q} \equiv \mathbf{c}\mathbf{b}^{-1}\mathbf{a} - \mathbf{a}\mathbf{b}^{-1}\mathbf{c}.$$

But
$$\begin{aligned}\mathbf{q}' &= \mathbf{a}'[\mathbf{b}^{-1}]'\mathbf{c}' - \mathbf{c}'[\mathbf{b}^{-1}]'\mathbf{a}' \\ &= -\mathbf{a}\mathbf{b}^{-1}\mathbf{c} + \mathbf{c}\mathbf{b}^{-1}\mathbf{a} \\ &= \mathbf{q}.\end{aligned}$$

The equation $$(\mathbf{x}'\mathbf{q}\mathbf{x}) = 0 \quad (\mathbf{q}' = \mathbf{q})$$

therefore represents a quadric, and the point $\boldsymbol{\alpha}$ lies on it, since $(\boldsymbol{\alpha}'\mathbf{q}\boldsymbol{\alpha}) = 0$. Hence the points of any line common to the three linear complexes lie on the quadric; in other words, the lines common to the three complexes generate the quadric.

It may make the last part of the argument clearer if we point out that the particular matrix $\mathbf{c}\mathbf{b}^{-1}\mathbf{a} - \mathbf{a}\mathbf{b}^{-1}\mathbf{c}$ was chosen precisely because it *is* symmetrical.

An alternative, though similar, ending could also be followed to show that *the lines common to the three linear complexes* $\mathbf{a}$, $\mathbf{b}$, $\mathbf{c}$ *also belong to the linear complex* $\mathbf{r}$, *where*

$$\mathbf{r} \equiv \mathbf{c}\mathbf{b}^{-1}\mathbf{a} + \mathbf{a}\mathbf{b}^{-1}\mathbf{c} \qquad (\mathbf{r}' = -\mathbf{r}).$$

ILLUSTRATION 5. *Show that the tangential equation of the pair of points in which a line* $\mathbf{p}$ *meets the non-singular quadric* $(\mathbf{x}'\mathbf{a}\mathbf{x}) = 0$ *is* $(\mathbf{u}'\mathbf{p}'\mathbf{a}\mathbf{p}\mathbf{u}) = 0$, *and that the coordinates of the tangent planes to the quadric at these points are given by*

$$\mathbf{a}\mathbf{p}\mathbf{u} = \pm\theta\mathbf{u},$$

where θ *is the square root of a polynomial in* a_{ij}, p_{ij}. [M.T. II.]

Let $\mathbf{y}$ be one of the points where the line meets the quadric and $\mathbf{u}$ an arbitrary plane through $\mathbf{y}$. Since the line meets the plane in the point $\mathbf{p}\mathbf{u}$, this is the same as $\mathbf{y}$, so that

$$\mathbf{y} = k\mathbf{p}\mathbf{u},$$

where k is a scalar. But $\mathbf{y}$ lies on the quadric, so that

$$(\mathbf{y}'\mathbf{a}\mathbf{y}) = 0,$$

or, cancelling k^2,

$$(\mathbf{u}'\mathbf{p}'\mathbf{a}\mathbf{p}\mathbf{u}) = 0.$$

This equation, satisfied by the coordinates of any plane through either of the two points, is the tangential equation of the point pair.

Now take $\mathbf{u}$ as the particular plane through $\mathbf{y}$ which touches the quadric, so that

$$\mathbf{u} = \mathbf{a}\mathbf{y}.$$

The line $\mathbf{p}$ meets this plane in the point $\mathbf{p}\mathbf{a}\mathbf{y}$, which must also be $\mathbf{y}$ itself. Hence

$$\mathbf{p}\mathbf{a}\mathbf{y} = \phi\mathbf{y},$$

where ϕ is a scalar.

Multiply each side on the left by $\mathbf{a}$ and use the relation $\mathbf{a}\mathbf{y} = \mathbf{u}$. Hence

$$\mathbf{a}\mathbf{p}\mathbf{u} = \phi\mathbf{u}.$$

In order to evaluate ϕ, write this equation in the form

$$[\mathbf{p} - \phi\mathbf{a}^{-1}]\,\mathbf{u} = \mathbf{0}.$$

This is, when written out, a set of four equations for the three ratios of the four unknowns u_i; eliminating u_i, we obtain the determinantal equation for ϕ

$$|\mathbf{p} - \phi\mathbf{a}^{-1}| = 0.$$

On expansion this is a quartic for ϕ.

Now a determinant is unaltered if rows and columns are transposed, so that the equation

$$|\mathbf{p}' - \phi[\mathbf{a}^{-1}]'| = 0$$

has the same roots. But $\mathbf{p}' = -\mathbf{p}$ since $\mathbf{p}$ is skew-symmetric, and $[\mathbf{a}^{-1}]' = \mathbf{a}^{-1}$, since $\mathbf{a}$ is symmetric; and so the latter equation is

$$|-\mathbf{p}-\phi\mathbf{a}^{-1}| = 0,$$

or

$$|\mathbf{p}+\phi\mathbf{a}^{-1}| = 0.$$

Hence if ϕ satisfies the quartic equation, so also does $-\phi$. The equation is therefore of the form

$$\alpha\phi^4+\beta\phi^2+\gamma = 0.$$

Moreover, $\phi = 0$ is a solution, since $|\mathbf{p}| = 0$ so that $\gamma = 0$. Hence the equation has two zero roots, which are irrelevant since $\phi = 0$ would lead to $\mathbf{u} = 0$. The equation for ϕ is therefore of the form

$$\alpha\phi^2+\beta = 0.$$

To express this equation with a unit coefficient for ϕ (so obtaining ϕ as a *polynomial*), we go back to the original equation, writing it in the form

$$(\mathbf{ap}-\phi\mathbf{I})\,\mathbf{u} = \mathbf{0}.$$

On elimination of $\mathbf{u}$, we obtain the equation

$$|\mathbf{ap}-\phi\mathbf{I}| = 0$$

for ϕ, and this, on expansion, is of the form

$$\phi^4+\alpha'\phi^3+\beta'\phi^2+\gamma'\phi+\delta' = 0,$$

where the coefficients are polynomials in the elements of **a and p.** But we have seen that this equation must have equal and opposite roots, two zero, so that $\alpha' = \gamma' = \delta' = 0$. Hence there remains the equation

$$\phi^2 = -\beta',$$

where $-\beta'$ is a polynomial in a_{ij}, p_{ij}. Writing $-\beta' = \epsilon$ (the negative sign does not matter, since we are working with complex coordinates), we have

$$\phi^2 = \epsilon$$

or

$$\phi = \pm\sqrt{\epsilon},$$

so that

$$\mathbf{apu} = \pm\sqrt{\epsilon}\,\mathbf{u},$$

where ϵ is a polynomial in a_{ij}, p_{ij}.

GENERAL EXAMPLES

1. O is a fixed point and ϖ a fixed plane. A variable point P is joined to O and P' is taken on OP so that P, P' are harmonically conjugate with respect to O and the intersection of P with ϖ. Prove that, if P describes a plane α, P' describes another plane α' which meets α in a line lying in ϖ.

The cone joining O to the section of the quadric S by a fixed plane α meets S again in points of a plane α'. If O describes a plane β, prove that the locus of the pole of α' with respect to S is a quadric touching S along its intersection with β. [P.]

2. Given a quadric Σ, show how to find sets of four points A, B, C, D such that each is the pole with regard to Σ of the plane containing the other three. Prove that, if the coordinates of the four points are (x_r, y_r, z_r, t_r) $(r = 1, 2, 3, 4)$, then the tangential equation of Σ may be written in the form

$$\sum_1^\lambda \alpha_r(x_r l + y_r m + z_r n + t_r p)^2 = 0.$$

Deduce, or prove otherwise, that, if A, B, C, D and A', B', C', D' are two such self-polar tetrads with regard to a quadric Σ, then every quadric which passes through seven of the eight points necessarily passes through the eighth. [P.]

3. Prove that a plane which touches both the quadrics

$$x^2+y^2+z^2+t^2 = 0, \quad x^2/a+y^2/b+z^2/c+t^2/d = 0$$

also touches the quadric

$$\frac{x^2}{a+\lambda}+\frac{y^2}{b+\lambda}+\frac{z^2}{c+\lambda}+\frac{t^2}{d+\lambda} = 0,$$

the locus of the points of contact, for varying λ, being a straight line.

Prove further that the straight lines form a ruled surface of order 8, which touches the quadric

$$x^2/a+y^2/b+z^2/c+t^2/d = 0$$

along its intersection with the quadric

$$x^2/a^2+y^2/b^2+z^2/c^2+t^2/d^2 = 0$$

and meets it again in eight generators. [P.]

4. Obtain the general equation of a quadric which contains the curve

$$x:y:z:t = \lambda^3:\lambda^2:\lambda:\lambda^4+\alpha,$$

where λ is a parameter and α a non-zero constant, showing that there is a pencil of such quadrics. Prove that each quadric contains four tangents of the curve and that the points of contact of the four tangents lie in a plane which, varying as the quadric varies in the pencil, passes through a fixed line. [P.]

5. Prove that all the quadrics which pass through seven points of general position in space also pass through an eighth point determined by these seven, and that the eight points are such that the cubic curve which passes through any six of them has the line joining the other two as chord.

There are twenty-eight lines which join pairs of the eight points, and each join is met in two points by the cubic curve through the other six points. Prove that the fifty-six points so determined all lie on the cubic surface which is the locus of the poles of a given plane with respect to the quadrics. Prove also that this locus meets each of the twenty-eight joins in the point which is the harmonic conjugate, with respect to the points joined, of the point where the join meets the given plane. [P., modified.]

6. AD, BE, CF are three skew lines in space of three dimensions. Show that the equation of the quadric of which they are generators can be expressed in the form

$$(adcx)(befx) = (adfx)(becx),$$

where $(pqrx)$ denotes the determinant whose columns are the coordinate vectors of the points P, Q, R and a variable point X.

If A, B, C, D, E, F are six points in general position in space, prove that the three quadrics defined by the triads of generators AD, BE, CF; AE, BF, CD; AF, BD, CE belong to a pencil. [P.]

7. Each of the equations $\delta_i = 0$ in tangential coordinates in space represents a point-pair $H_i H_i'$. Show that the quadrics of the tangential pencil $\delta_1 + \lambda\delta_2 = 0$ have four common generators.

Show that if there is a set of constants k_i for which the identity

$$I \equiv k_1\delta_1 + k_2\delta_2 + k_3\delta_3 + k_4\delta_4 \equiv 0$$

holds, then in general the lines H_1H_2, $H_1'H_2'$, H_3H_4, $H_3'H_4'$, H_1H_2', H_2H_1', H_3H_4', H_4H_3' lie on a quadric, four belonging to each system of generators, and that the same is true with any permutation of the suffixes.

Find the configurations of the set of eight points

(i) when $k_4 = 0$,

(ii) when there is an identity $k_1\delta_1 + k_2\delta_2 \equiv \delta_0$, in addition to $I \equiv 0$. [M.T. II.]

8. Prove that the chord of the twisted cubic $(\lambda^3, \lambda^2\mu, \lambda\mu^2, \mu^3)$—referred to a homogeneous parameter λ/μ—through the point $P(\xi, \eta, \zeta, \tau)$ meets the curve in the two points A, B whose parameters λ/μ are the roots of the quadratic

$$H \equiv (\lambda\eta - \mu\xi)(\lambda\tau - \mu\zeta) - (\lambda\zeta - \mu\eta)^2 = 0.$$

Thence, or otherwise, obtain the equation of the ruled surface of tangents to the curve.

Show further that, if the Jacobian of H and the cubic form

$$f(\lambda, \mu) \equiv \tau\lambda^3 - 3\zeta\lambda^2\mu + 3\eta\lambda\mu^2 - \xi\mu^3$$

is $\delta\lambda^3 - 3\gamma\lambda^2\mu + 3\beta\lambda\mu^2 - \alpha\mu^3$, then $(\alpha, \beta, \gamma, \delta)$ is the harmonic conjugate of P with regard to A, B. [M.T. II.]

9. Given a quadric, prove that, if the lines joining the vertices of one tetrahedron to the poles of the faces of another tetrahedron meet in a point, then the lines joining the vertices of the second tetrahedron to the poles of the faces of the first also meet in a point. [M.T. II, Schedule B.]

10. Prove that the four planes joining four points of a twisted cubic to any chord are projective with the four planes joining these four points to any other chord.

Considering four points of the curve and the four planes containing the triads of these points, let these planes meet an arbitrary plane in four lines. Prove that the seven lines, constituted by these and the three chords of the cubic curve which lie in this plane, touch a conic.

Prove further that, if three tetrahedra are inscribed in the curve, the twelve faces of these touch a quadric. [M.T. II, Schedule B.]

11. Show that the equations

$$x:y:z:t = (\theta-\alpha)^{-1}:(\theta-\beta)^{-1}:(\theta-\gamma)^{-1}:(\theta-\delta)^{-1},$$

where θ is a variable parameter, represent a twisted cubic curve passing through the vertices of the tetrahedron of reference and through the point $(1, 1, 1, 1)$, and that any chord of the curve meets the faces of the tetrahedron in four points whose cross-ratio is constant.

Prove that a fixed plane $ax+by+cz+dt = 0$ is met by any cubic curve through these five points in three points forming a triad self-polar with respect to the conic in which the plane meets the quadric

$$ax^2+by^2+cz^2+dt^2 = 0. \qquad \text{[M.T. II.]}$$

12. Prove that if $P_i = 0\ (i = 1, \ldots, 8)$ are the (tangential) equations of eight points, the *necessary and sufficient* condition that they should be associated, i.e. that the quadrics through any seven of the eight points all pass through the eighth, is that there should be an identical relation of the form $\sum_{i=1}^{8} \lambda_i P_i^2 \equiv 0$, where $\lambda_1, \ldots, \lambda_8$ are constants all different from zero.

Show that the eight vertices of two tetrahedra self-conjugate with regard to a quadric are associated, and also that if a set of eight associated points is divided into two sets of four points the eight faces of the two tetrahedra so formed are associated in the dual of the above sense. [M.T. II.]

13. Prove that a line meets its polar line with regard to a given quadric S if, and only if, the line is a tangent of S.

By considering the effect of reciprocation on the generators of the quadric S', or otherwise, show that, if S' is reciprocated into itself with regard to S, then S and S' either touch along a conic or have four generators in common.

Show that in the former case the equations of S and S' can be written

$$x^2+y^2+z^2+t^2 = 0, \quad x^2+y^2+z^2-t^2 = 0,$$

and in the latter case the equations can be written

$$yz = xt, \quad yz+xt = 0. \qquad \text{[M.T. II.]}$$

14. With projective coordinates in space the points (x, y, z, t) and planes (l, m, n, p) correspond according to a general (non-singular) linear formula

$$\lambda l = a_1x+a_2y+a_3z+a_4t,$$

$$\lambda m = b_1x+b_2y+b_3z+b_4t,$$

$$\lambda n = c_1x+c_2y+c_3z+c_4t,$$

$$\lambda p = d_1x+d_2y+d_3z+d_4t.$$

Show that the points which lie on their corresponding planes are the points of a quadric and the planes which pass through their corresponding points touch another quadric.

To the point P corresponds the plane π and to the points of the plane π' correspond the planes through P. Show that if P lies on a fixed line then the intersection of π and π' belongs to one system of generators of a quadric, and that if P lies on a fixed plane the intersection of π and π' is a chord of a certain twisted cubic. [M.T. II.]

15. Prove that the polar planes of a point P with respect to the quadrics of a 'net'

$$\lambda_1S_1+\lambda_2S_2+\lambda_3S_3 = 0,$$

where S_1, S_2, S_3 are quadrics not belonging to a pencil, are in general concurrent in a point P', and that as P describes a line l the point P' describes a space cubic curve of which the chords are the polar lines of l with respect to quadrics of the net.

Show that, if l is the join of two of the eight points common to all quadrics of the net, the cubic curve breaks up into l and two other lines meeting l. [M.T. II.]

16. Prove that the locus of the vertex of a cone which touches the sides of a skew hexagon consists of the planes containing two consecutive sides of the hexagon and a non-singular quadric. In what circumstances does the quadric degenerate?

Show that the locus is the whole space only if the hexagon lies on a quadric, or is a plane hexagon circumscribing a conic. [M.T. II.]

17. A, B, C are three points generally situated with respect to a quadric Ω. Prove that the planes joining A, B, C to the polar lines of BC, CA, AB with respect to Ω meet in a line λ, and show that, if D is a fourth point, the tetrahedron $ABCD$ is in perspective with its polar tetrahedron with respect to Ω *if, and only if*, D lies on λ. [M.T. II.]

18. A twisted cubic Γ is given parametrically in the form $(\theta^3, \theta^2, \theta, 1)$, and a linear complex K has the equation

$$a'l+b'm+c'n+al'+bm'+cn' = 0.$$

Prove that the planes which join the chords of Γ belonging to K to the point on Γ whose parameter is σ envelop the cone

$$\begin{vmatrix} 2c' & -b' & a-a' & x-\sigma y \\ -b' & 2a' & b & y-\sigma z \\ a-a' & b & 2c & z-\sigma t \\ x-\sigma y & y-\sigma z & z-\sigma t & 0 \end{vmatrix} = 0.$$

Deduce that the intersections (apart from its vertex) of this cone with Γ are independent of σ, and show that they are the points such that the two chords of Γ which pass through them and belong to K coincide. [M.T. II.]

19. The coordinates, referred to a tetrahedron $ABCD$, of four points A', B', C', D' are given by the respective four columns of a non-singular skew-symmetrical matrix of order four. Show that the tetrahedron $A'B'C'D'$ is both inscribed and circumscribed to $ABCD$.

Verify that there is a linear identity connecting the equations of the four point-pairs AA', BB', CC', DD'; and hence show that any one of these four consists of a pair of conjugate points for the quadric which contains the joins of the remaining three pairs.

Deduce that the two transversals of the four joins cut each join harmonically. [M.T. II.]

20. Prove that, if A, B, C are three points on a conic, the lines which join them to the vertices of the triangle formed by the tangents to the conic at A, B, C meet in a point O.

A quadric Q circumscribes a tetrahedron T, so that each face of T meets Q in a conic containing three vertices of T from which there arises a point O as in the first part of the question. Prove that, if the tangent plane of Q at any one vertex of T passes through the point O belonging to the opposite face of T, then the tangent planes at all four vertices have this property.

Prove also that, independently of the above condition, if the four points O, one in each face of T, are coplanar, then Q is a cone, and conversely. [M.T. II.]

21. Γ is a twisted cubic; P a point in space. Prove that, in general, there are just three points of Γ at which the osculating planes contain P, and that the plane π of these points also contains P. Show that there is a polarity (null system) N in which the polar plane of P is always π. Show also that the polar planes of the points of a line λ have in common a line λ', and identify λ' when λ is a chord of Γ.

A, B are two points of Γ; l is the line common to the osculating planes at A and B. Show that the polar lines of l with respect to those quadrics which contain Γ and its chord AB generate a quadric Q.

Obtain, by considering the reciprocals of these quadrics with respect to N, a geometrical definition of the quadric Q' into which Q is reciprocated by N, and show that Q and Q' intersect in a skew quadrilateral. [M.T. II.]

22. Prove that the quadrics upon which a given twisted cubic curve k lies constitute a net; and show that any conic which meets k at three points lies on just one quadric through k.

Two given conics s, t in different planes have each three points on k, these six points being all different. Show that, in general, there is just one chord of k which meets s and t at points not on k; and discuss the special case when s, t have two points in common, not on k. [L.]

23. The coordinates of the join of (x_1, y_1, z_1, t_1) and (x_2, y_2, z_2, t_2) are defined by $l = t_1x_2 - t_2x_1$, $l' = y_1z_2 - y_2z_1$. [This is not the notation of the text; the alternative form is, however, retained for practice.] Three lines of general position in space are $(l_i, m_i, n_i, l'_i, m'_i, n'_i)$ for $i = 1, 2, 3$. Show that the determinant in

$$| l'_i t + m_i z - n_i y,\ m'_i t + n_i x - l_i z,\ n'_i t + l_i y - m_i x | = 0$$

is divisible by t and that, when this factor is removed, the resulting equation $\mathscr{L}$ is the equation, as a locus in point coordinates (x, y, z, t), of the quadric Q which contains the lines. Write down the determinant which similarly gives rise to the equation $\mathscr{E}$, as an envelope in plane coordinates, of Q.

If two of the three lines meet, find, by taking these lines as two edges of the tetrahedron of reference, what quadric is then represented by $\mathscr{L}$.

It is seen from $\mathscr{L}$ and $\mathscr{E}$ that the equation of Q, whether in point or plane coordinates, is homogeneous and of degree 1 in the six coordinates of any one of the three lines. Yet the coefficients in either the point or the plane equation of a quadric are of degree 3 in those of the other equation. Explain this apparent paradox. [M.T. II.]

ANSWERS TO EXAMPLES

Examples I

4. Coplanar.

Examples II

4. $(a+b+c+d)(ax^2+by^2+cz^2+dt^2) = (ax+by+cz+dt)^2$.

16. $(0, 1, 2, 3)$, $(1, 0, -1, -2)$, $(2, 1, 0, -1)$, $(3, 2, 1, 0)$;
$3x+y+2z+3t=0$, $x+z+t=0$, $x+y+t=0$, $3x+2y+z+3t=0$.

Miscellaneous Examples II

3. $a^2x^2+b^2y^2+c^2z^2+d^2t^2=0$.

Examples III

1. $4x+3y-z-2t=0$.

Miscellaneous Examples III

4. Quadric.
5. $\Sigma ax^2+abct^2+\Sigma(b+c)\,yz+\Sigma a(b+c)\,xt=0$.
9. $(\mu_1\,\mu_2\,\mu_3\,\mu_4)$

Examples IV

1. $(3, 2, 1, 1, -2, 1)$; $(5, 2, 1, 1, -2, -1)$.
2. $(-2, 1, 1, -2, -3, -1)$; $(0, 1, -3, 2)$, $(-1, 0, 2, -1)$, $(3, -2, 0, -1)$, $(-2, 1, 1, 0)$.
3. $(0, -5, -9, -3, 9, -5)$.
4. $(2, 1, -2, 1, 2, 2)$; $2x-6y-z+5t=0$.

Miscellaneous Examples IV

9. $x^2-y^2-yz-zx+xt-yt+2zt=0$.

Examples V

1. $x-3y+3z-t=0$, $x-6y+12z-8t=0$, $x-9y+27z-27t=0$.
2. $(3, 2, 1, 1, -2, 1)$.

Miscellaneous Examples VI

1. Meet in two conics, in planes $\alpha=0$, $\beta=0$.

$$ax^2+by^2+2hxy+2wzt=0,\quad c'z^2+d't^2+2h'xy+2w'zt=0.$$

7. Line $x+ay=0=z+at$; line $bx+z=0=by+t$; conic $ax-bt=0=xt-yz$.

General Examples

4. $(x^2-yt+\alpha z^2)+\lambda(y^2-zx)=0$.
7. (i) Six vertices of plane quadrilateral, and two arbitrary points,
(ii) two quadrilaterals with a pair of opposite vertices in common.
8. Ruled surface: see p. 89, Ex. 12.

23. If $\mathscr{F}=0$ is the 'ordinary' envelope equation of Q, then $\mathscr{F}\equiv 0$ when any two of p_1, p_2, p_3 meet. Hence $\mathscr{F}$ contains $\varpi_{23}\varpi_{31}\varpi_{12}$ as a factor, and this factor is of degree 2 in the coordinates of each of p_1, p_2, p_3.

INDEX